The Reluctant
Mr. Darwin

FORTHCOMING TITLES

Richard Reeves on Rutherford and the Atom

Daniel Mendelsohn on Archimedes and the Science of
the Ancient Greeks

Lawrence Krauss on the Science of Richard Feynman

General Editors: Edwin Barber and Jesse Cohen

GREAT DISCOVERIES

DAVID QUAMMEN

The Reluctant Mr. Darwin

An Intimate Portrait of Charles Darwin and the Making of His Theory of Evolution

ATLAS BOOKS

W. W. NORTON & COMPANY
NEW YORK · LONDON

Small portions of this book, in slightly different form, appeared previously in the author's 2001 Bradley Lecture at the Library of Congress and in his article "Was Darwin Wrong?" in the November 2004 issue of *National Geographic*.

Manufacturing by R.R. Donnelley, Bloomsburg Division
Book design by Chris Welch
Production manager: Julia Druskin

Library of Congress Cataloging-in-Publication Data

Quammen, David, 1948–

The reluctant Mr. Darwin : an intimate portrait of Charles Darwin and the making of his theory of evolution / David Quammen. — 1st ed.

p. cm. — (Great discoveries)

Includes bibliographical references and index.

ISBN-13: 978-0-393-05981-6 (hardcover)

ISBN-10: 0-393-05981-2 (hardcover)

1. Darwin, Charles, 1809–1882. 2. Natural selection. 3. Naturalists—England—Biography. I. Title. II. Series.

QH31.D2Q35 2006

576.8'2092—dc22

2006009864

Atlas Books, LLC, 10 E. 53rd Street, New York, N.Y. 10110

W. W. Norton & Company, Inc., 500 Fifth Avenue, New York, N.Y. 10110
www.wwnorton.com

W. W. Norton & Company Ltd., Castle House, 75/76 Wells Street, London W1T 3QT

1 2 3 4 5 6 7 8 9 0

To Betsy

Contents

Home and Dry

an introduction

Charles Darwin holds a peculiar position in the history of science and society. His name is a household word but his ideas—with a single exception—aren't household ideas. He's central, he's iconic, but that's not to say that he's widely and well understood. If the scientific community issued bank notes, true enough, the face on the dollar bill would be Darwin's. It's a good face, an amiably stolid face, like George Washington's as engraved from the painting by Gilbert Stuart; yet it conceals, like Washington's, deep veins of complexity and tension. Everyone knows something about who Darwin was, what he did, what he said, and the thing that most people think they know is: He concocted "the theory of evolution." This isn't quite wrong, just confused and imprecise, but it misses those points about Darwin's work that are most profoundly original, and dangerous, and thrilling.

Both as hero and as bugaboo, Darwin is taken for granted in a way that Copernicus, Kepler, Newton, Linnaeus, Charles Lyell, Gregor Mendel, Albert Einstein, Marie Curie, Niels Bohr, Werner Heisenberg, Alfred Wegener, Frederick Hubble, James Watson, and Francis Crick are not. One measure of his supposed familiarity is the careless use, within common discourse, of the terms "Darwinism" and "Darwinian," which presume at reducing to trademark simplicity a diverse body of work that can't be so easily reduced. Forget about Darwinism, it doesn't exist. Not unless you define it by arbitrary stipulation—these concepts included, those concepts not—in a way that Darwin himself never did. And what is Darwinian? Well, a fascination with fancy pigeons is Darwinian, in the sense that our man, during one period, became entranced by his aviary full of pouters and fantails and runts. A fondness for long solitary strolls, not far from home, is Darwinian. Recurrent bouts of unexplained vomiting are, as you'll see, very Darwinian. My point is this: Charles Darwin didn't found a movement or a religion. He never assembled a creed of scientific axioms and chiseled them onto a stone tablet beneath his own name. He was a reclusive biologist who wrote books. Sometimes he made mistakes. Sometimes he changed his mind. Sometimes he worked on little subjects and sometimes on big ones. True, most of his published writings share a single underlying theme—the unity of all life, reflecting the processes of evolution. But he particularized that theme in a variety of concepts, some of which interlock nicely and remain valuable to biology, some of which don't. It's better to examine his ideas individually than to try to bundle them as a brand.

Copernicus, among the great scientists mentioned above, is the one whose impact most closely resembles Darwin's, in that Darwin continued the revolution Copernicus began, alerting humans to the fact that we don't occupy central position in the universe. Darwin extended that recognition from cosmology to biology. "People often talk," he muttered to himself, in an early notebook, "of the wonderful event of intellectual Man appearing." Darwin, for his part, wasn't so impressed by the emergence of "intellectual Man," adding contrarily that "the appearance of insects with other senses is more wonderful." This heretical comment shows that, from the start of his musings about how species originate, Darwin denied mankind its self-assigned demigod status and included us in the jumble of struggle and change. He was no humanist (though he was always humane). For sheer wonder, give him not the brain of *Homo sapiens* but the orienteering and architectural instincts of the honey bee.

I say Darwin "continued" rather than "completed" the Copernican revolution against anthropocentrism because the battle is still going on. Many people, even among those who would say they accept Darwin's theory of evolution (whatever they take it to be), decline to absorb the full implications of what he wrote. His biggest idea, bigger than mere evolution, was just too big, too harsh and threatening. That idea was what he called "natural selection" and identified as the primary mechanism of evolutionary change. According to Darwin's view (since reaffirmed by a century and a half of further biological evidence), natural selection is a purposeless process but an efficacious one. Impersonal, blind to the future, it has no goals, only results. Its sole standards of valuation are survival and reproductive success. From scattershot variations, culled and accreted, it produces pragmatic forms of order. Its driving factors are hyperfecundity and mortal competition; its products and byproducts are adaptation, complexity, and diversity. It embodies a deep chanciness that is contradictory to the notion that Earth's living creatures, their capacities (including human capacities), their histories, their indigenousness to particular locales, and their interrelations all reflect some sort of divinely preordained plan. Creationist proselytizers pursuing Christian political agendas are therefore right to regard it with loathing and alarm.

Those creationist proselytizers aren't alone in their dissent from evolutionary thinking. They've had reason to feel encouraged, within recent years, by the high level of lingering resistance—at least in the United States—to what Darwin articulated back in 1859. Their political challenges (within various state legislatures and local school boards) have been persistent but mostly unsuccessful. Important court cases (such as *Edwards v. Aguillard,* in 1987, wherein the U.S. Supreme Court declared Louisiana's creationism-in-the-schools law unconstitutional, and *Kitzmiller v. Dover,* in 2005) have gone against them. But they're correct about one thing: Broader public opinion harbors a startling level of ambivalence on this subject. Postmodern America is a hotbed of pre-evolutionary views.

You may have heard loose assertions to the effect that a third of all Americans—or is it 40 percent, or more?—don't accept the reality of evolution. Here are some hard numbers: 45, 47, 44. The Gallup organization, in November 2004, after more than a thousand telephone interviews, found that 45 percent of their respondents agreed with the statement: "God created human beings pretty much in their present form at one time within the last 10,000 years or so." For short: creationism. Another statement, offered alternatively, said that humans "have developed over millions of years from less advanced forms of life, but God guided this process." For short: theistic evolution. That option satisfied 38 percent of the people polled. Just 13 percent agreed with a statement that humans have developed from other life forms *without* guidance by God. For short: materialistic evolution. (And the remainder had no classifiable opinion. For short: Go away, we're watching TV.)

The most striking thing about these poll results is not that resistance to evolutionary theory registered so high in one poll or another; the most striking thing is that it remained virtually unchanged in six parallel samplings over the course of a generation. Back in 1982, presenting the very same options, Gallup found 44 percent of respondents agreeing that God, not evolution, had created human beings. In 1999, the percentage peaked at 47, and it has never fallen lower than 44. If these polls can be trusted, almost half the American populace chooses to understand the origin of our species as though Charles Darwin never lived. Another major increment, ranging between 37 and 40 percent over the years, prefers the "guided by God" option, theistic evolution, which is still utterly contrary to what Darwin proposed. Summarizing the arithmetic: Between 81 percent and 87 percent of Americans reject Darwin's view of human evolution.

Gallup isn't alone in measuring this phenomenon. A more recent poll, conducted in July 2005 by the Pew Research Center for the People and the Press, along with a partner organization, found 42 percent (among 2,000 Americans interviewed) affirming that "living things have existed in their present form since the beginning of time." Another 18 percent subscribed to theistic evolution, at least with respect to humans, specifying that the process must have been "guided by a supreme being." So the Pew results are slightly less negative in total than Gallup's: only a 60 percent rejection of Charles Darwin, instead of 80-some.

Maybe the polls are invalid. Maybe the numbers would be

much different in England or Sweden or India. Maybe the same distinctly American mixture of skepticism and evangelicism that led to the Scopes trial, in 1925, continues to animate many citizens who would simply rather take their biology from scripture than from science. Maybe the question of human evolution is misleading and inordinately touchy; maybe Gallup and Pew should be asking whether God created, let's say, *tree kangaroos* in their present form. Or maybe . . . who knows? I don't claim to have any definitive explanation for such an extreme level of skepticism and willful antipathy toward such a well-established scientific discovery. Frankly, it mystifies me. But certainly those Gallup results—combined with the continuing political offensive against teaching evolutionary biology in public schools—testify that Charles Darwin isn't just perennially significant. He's also urgently relevant to education and governance.

Speaking personally for a moment: I come to the subject by a roundabout route. I'm not a biologist. I'm not a historian. I have virtually no academic training in science. Nevertheless, for the past twenty-five years, I've made my living primarily as a science journalist, learning what evolutionary biology and ecology I know by self-education (that is, reading, especially of scientific journals) and pestiferous questioning of experts. During those years, I've had a privileged sort of opportunity: much field time with field biologists. On assignment to various magazines, and while doing research for books, I've been welcomed to tramp through tropical forests, ascend rivers from Mongolia to the Amazon, stroll across equatorial savannahs, prowl remote islands, and otherwise knock around outdoors with some of the world's brightest and hardiest natural scientists. Besides advancing (slowly) my understanding of certain ecosystems and species, and of some of the underpinning concepts of ecology and evolutionary biology, these experiences have shown me that field biologists are, on the whole, a guild of extraordinary people—smart, passionate, patient, congenial, and physically as well as intellectually tough. Some people admire soldiers, or surgeons, or firemen, or astrophysicists, or medical missionaries, or cowboys. I admire field biologists.

This is part of what brings me to Darwin. He himself was a field biologist, of course, during one crucial period of his life: the four years, nine months, and five days he spent as a naturalist aboard the *Beagle*, a British naval ship sent out to chart certain stretches of South American coastline. That voyage lasted from 1831 until 1836. Darwin was in his mid-twenties, just the right age for maximum exertion in difficult circumstances and maximum absorption of new facts and impressions. While the *Beagle*'s captain and crew did their work, young Mr. Darwin collected marine specimens with a plankton net dragged behind the ship and made long excursions ashore for further collecting and observing. Inexperienced at the start, he gradually became a methodical and keenly percipient scientist. He visited Brazil, Uruguay, Argentina, Chile, Peru, New Zealand, Australia, South Africa, and a number of small oceanic islands, including the Cape Verdes, the Azores, Tahiti, Mauritius, St. Helena, and the Galápagos. Landing at Falmouth in southwestern England on October 2, 1836, he would never leave Great Britain again. His days of gallivanting field biology were over. He was home and dry, and quite happy to remain so, at least for a while. Other biologists of his era (such as Alfred Russel Wallace and Henry Walter Bates, about whom more below) might spend a decade at punishing fieldwork in the Amazon or Borneo or wherever; but for Darwin, five years was a bellyful. Most of his scientific labor, throughout the rest of his life, would entail research reading, correspondence, experimentation, dissection, observation in the meadows and woodlands not far from home, and thinking. Partly because of health problems, partly by intellectual disposition, he became largely an indoor guy.

Indoors was where he developed his ideas. So, notwithstanding my bias toward field biologists, and the importance of those vivid early experiences in fueling Darwin's later thought, I've made a counterintuitive choice: to omit the voyage (except as background) from this account, and to take up the story just afterward. Why ignore the most famous episode of Darwin's life? Three reasons. First, because it *is* the most famous. Whatever else you may know about Charles Darwin, you probably know that he once sailed on a ship called the *Beagle*, visited the Galápagos Islands, and saw there some interesting reptiles and birds. My second reason is a matter of economy and scope. To say it more plainly: brevity. Darwin's life story has been told many times, by some excellent biographers (notably Janet Browne, in her magisterial two-volume *Charles Darwin*, and the team of Adrian Desmond and James Moore, in their trenchant 800-page *Darwin: The Life of a Tormented Evolutionist*) and by some less excellent ones, but most people haven't read that story even once. Of course in each telling it's a slightly different story, depending on selection, omission, and the biases and purposes of the teller. My purpose has been to create a concise treatment, part narrative and part essay, accurate but pleasantly readable, of this huge and deeply complicated subject. I wanted to sketch, in not many pages, the growth and development of a man of ideas, with particular focus on just one of them. Third reason for skipping the *Beagle* years: Darwin's later intellectual adventures are, in my opinion, even more exciting than the romps across Patagonia and the Galápagos.

Chief among those adventures is the discovery of natural selection. That idea, taken freshly, with its full implications, is marvelous and shocking and grim. It's even more marvelous when you consider its provenance: a deeply radical insight from a deeply cautious man. The shy patriarch with the bald head and the full beard, the breeder of pigeons and primroses, the very private Englishman who wound up buried in Westminster Abbey, the fellow with a good face for bank notes, presents to us a comfortably dowdy image; but not everything about Charles Darwin is so comfortable. At the core of his work is a difficult, scary materialism. That's one of the themes I try to explore in this book. Another is that it was difficult and scary even to him.

The Fabric Falls

1837–1839

1

In the early weeks of 1837, Charles Darwin was a busy young man living in London. Ambitious, intellectually awakened from a drowsy postadolescence, excited by opportunity, he was newly defining his life. He didn't yet recognize the awful scope of the idea that was growing inside him. On February 12, he turned twenty-eight.

Darwin had been home from his round-the-world journey aboard the survey ship *Beagle* just since the previous October. He was glad to be back on solid ground, walking floors that didn't tilt with the waves. During the course of the trip, which had originally been projected to last only two or three years but ended up stretching to half a decade, he had transformed himself drastically—from an unfocused graduate of divinity schooling at Cambridge, with a gentleman's passion for bird shooting and a collector's enthusiasm for rare beetles, into a serious student of geology and natural history. Even his widowered father, the grumpy and obese Dr. Robert Darwin, saw a difference. The doctor had once scolded him for being a feckless young sport, who cared only for bird-hunting and rat-catching and would be "a disgrace to yourself and all your family." But now Charles's reputation as a scientific traveler had preceded him home, and papa was placated. On first glimpse of his son after the voyage, Dr. Darwin turned to Charles's sisters and said, "Why, the shape of his head is quite altered." If it wasn't true phrenologically, it was apt as metaphor. The shape of his thinking had changed. It would soon change even more.

After a brief visit with his father and sisters at the family home in Shrewsbury (a medium-sized town up in Shropshire), and then a short stay in Cambridge near his old university chums, Darwin had come to the big city and rented rooms in a house on Great Marlborough Street, within walking distance of important scientific institutions such as the Zoological Society and the British Museum. He hated London, with its Dickensian smog and clatter, but he had purposes for tolerating it. His days were full of follow-up chores involving his scientific harvest from the *Beagle* voyage. That harvest included facts, notes, and ideas, but also mammal pelts, bird skins, pickled reptiles and fish, dried plants, and fossils. He had sent back crates, bottles, and casks of specimens from South America during the ship's years of surveying there, and had brought more with him on board, most of which were now farmed out to experts for identification and study. He had been a scientific nobody when he set sail on the *Beagle*, as unofficial naturalist (there was another naturalist, more official though less ardent, until that fellow resigned in a jealous huff) and social companion to the captain; but he'd proved himself vastly competent. His productive collecting in exotic locales and his sharply observant letters had given him some buzz in scientific circles even before he got home. He was considered a talented comer, and the uptake on his specimens was good. Richard Owen, a brilliant anatomist at the Royal College of Surgeons, had agreed to describe the fossil mammals. George Waterhouse, a museum curator, took on the living mammal species and the insects. John Gould, a respected ornithologist, would do the birds. Thomas Bell, a dentist turned zoology professor, got the reptiles. Darwin himself had meanwhile begun writing a book. This was a big step for him, implying a new level of confidence in his own observations and voice. A *book*, imagine. Yes, because he'd seen things that few other people ever had. He'd gathered impressions and data, carefully. It would be a pastiche of travel narrative, cultural portraiture, geology, and natural history derived from his diary of the voyage.

The unwritten book was already under contract to a publisher—as arranged by the *Beagle*'s captain, Robert FitzRoy, a very capable man but a cantankerous one, aristocratic and tippy. It had been FitzRoy's perfectionism, as well as complicating circumstances, that stretched a two-year voyage almost to five. Now the captain wanted a multi-volume record of his ship's recent expeditions, and he was glad to slot Darwin's book into the package; FitzRoy himself would do another volume, if he ever got around to it. Darwin set himself going and, energized by the prospect of becoming a published author, scribbled hard. The *Beagle* diary was his core material, but he wanted to add narrative flow, a few ideas, and some polish. He confided to a Cambridge friend, William Darwin Fox (who happened also to be a second cousin), his discovery that "writing is most tedious & difficult work." But he had one advantage that made the task easier: a sizable annual allowance from his father. He wasn't pressed to go looking, at least not yet, for a day job.

Socially he was in demand, as a returned traveler with stories to tell who happened also to be an eligible bachelor. For a while, that suited him fine. Charles Lyell, the rising star of English geologists, whose three-volume *Principles of Geology* was changing the way people thought about earth sciences, had welcomed him as a new friend and protégé. The inventor Charles Babbage began inviting him to fancy parties. Darwin's elder brother Erasmus, trained as a physician but with no desire to practice (and no need, thanks to their father's money), was already ensconced as a *bon vivant* in the city. Erasmus, hosting small gatherings at his own place on Great Marlborough Street, pulled Charles into a circle of bright people that included the political writer Harriet Martineau and the crusty Scottish historian Thomas Carlyle. Leonard Horner, an eminent educator and scientist, had a house full of unmarried daughters to which Darwin paid some flirtatious visits, though not as many as Mr. Horner would have liked. Darwin's five years on the *Beagle* had been lonesome, notwithstanding the meals with FitzRoy and the cramped cabin he'd shared with an officer and a midshipman, and during these first months in London he made up for it, basking in clever dinner-table chat, flattering attention, and female company. With Lyell vouching for him, he was elected to the Athenaeum Club (in the same group, speaking of Dickensian milieu, that included Charles Dickens himself), and that became his refuge for dining quietly and reading the journals. He attended meetings of the Zoological Society and the Geological Society, sometimes presenting a short paper himself. None of it kept him from progress on the book. He had taught himself discipline, as well as a lot of geology and biology, aboard the *Beagle*.

Just days after settling into London, he met with John Gould to talk about his bird specimens. Gould called his attention to a bunch that Darwin had collected in the Galápagos archipelago, six hundred miles off the west coast of South America, during the ship's brief stop there in late September and October 1835, on the way home by circumnavigation. They were all smallish and brownish, these birds, but with various shapes and sizes of beaks. Darwin had taken them to be a mixed assortment of wrens, grosbeaks, orioles, and finches, and hadn't bothered to label which ones came from which islands. The lack of labeling had been, in retrospect, a frustrating mistake. But as a field naturalist with wide-angle interests, uncommitted to any theory, he hadn't known just what he was looking for. In January 1837, four months after the *Beagle*'s return, he'd heard Gould deliver a preliminary report on the wren-grosbeak-oriole bunch at a Zoological Society meeting. It contained a surprise: They were all finches, Gould said. Big beaks and little beaks, sharp beaks and blunt beaks, there were a dozen species, closely related but distinct, representing some unfamiliar new group. Now, at the private consultation with Darwin in March, Gould went further: *thirteen* species of finches, all unknown to science. And not only that. Among another bunch, which Darwin had recognized as mockingbirds, Gould found three distinct mockingbird species. Unlike the finches, the mockingbirds had reached Gould with island-of-origin tags; because they were less diverse, less confusingly intermixed in the wild, Darwin had been more meticulous as he collected them. Funny thing about these mockingbirds, said Gould.

Each species, according to your labels, inhabits a different island.

That was weird, thrilling news. One species per island, all new? It confirmed something Darwin had whispered to himself in his ornithological notes while the *Beagle* was still at sea. Isn't it strange, he wrote, that these different kinds of bird, distinct but related, filling similar roles, live separately on closely neighboring islands? Maybe, contrary to received wisdom about the origin of all forms of life, they're just varieties derived from a common stock. Maybe they weren't *created* in the theological sense—that is, by a divine act of special creation for each. Maybe they just . . . happened. "If there is the slightest foundation for these remarks," Darwin told himself, and no one else, "the Zoology of Archipelagoes will be well worth examining; for such facts would undermine the stability of species."

He was more right than he knew. Species weren't stable, and islands held some of the best clues.

Other bits of unsettling data came to him, around the same time, in reports on his other specimens. Richard Owen had identified a giant extinct ground sloth, a giant extinct armadillo, and what he took to be a giant extinct capybara among Darwin's fossils from the South American mainland. It seemed oddly coincidental—to Darwin, if not to Owen—that such extinct forms should be found in the same geographical areas inhabited by living versions of sloth, capybara, and armadillo. John Gould announced at the next Zoological Society meeting, on March 14, that Mr. Darwin had discovered a new species of flightless bird, a smallish rhea—Gould named it *Rhea darwinii*—in southern Patagonia, just adjacent to the distributional range of the larger rhea, already known.

In the meantime Thomas Bell was finding island-by-island differences among the Galápagos iguanas. And now Darwin remembered something the vice-governor of the islands had told him about the giant tortoises: They too were distinguishable, island by island, from the shapes of their shells. Darwin put these facts together and asked himself *why*? Why should forms closely resembling one another, alive or extinct, be found clustered side by side?

It's not possible to say exactly when Charles Darwin became an evolutionist. He didn't blurt out his *Eureka!* in a letter, or a journal paper, or a fevered talk to one of the societies. At this point he was circumspect, uneasy, and mum. He had reason to be. England was a tumultuous place in the late 1830s, with a badly depressed economy, a new Poor Law replacing old-fashioned charity with grim workhouses, and a Chartist movement (named for its "People's Charter," a platform of working-class empowerment) staging mass protests to demand democratic reforms. Early evolutionary ideas about progressive change among species, as suggested by French zoologists such as Jean-Baptiste Lamarck and Etienne Geoffroy Saint-Hilaire, had been absorbed by English and Scottish radicals into their arguments for progressive social change, causing nervy discomfort to the Whigs who controlled Parliament and the Anglican prelates who ran the national Church, with all its wealth and other vested interests. And their discomfort couldn't be lightly ignored. Christianity as interpreted by Anglican leaders was not just the predominant religion in England; it was the *official* religion. The country hadn't had a revolution since 1688, and Chartism plus economic depression suggested that another might be imminent. Darwin, taking his first steps over the line between tradition and evolution, found himself occupying ground near those battle lines of class and religious warfare. He moved carefully. Didn't announce his apostasy. Still, it's possible to approximate the timing of this intellectual conversion: March of 1837, soon after his talks with Gould and Owen. Species changed, one into another. He knew it. He just didn't know how.

Months afterward he made another note, regarding the curious characteristics of his South American fossils and the Galápagos species he'd seen: "These facts origin (especially latter) of all my views." But for now, he was keeping those views to himself.

2

He didn't use the word "evolution," not until later, not for decades. In July of that year he began what he called his notebooks on the "transmutation" of species. The first of them was a pocket-size booklet bound in brown leather with a metal clasp, small enough to be carried in a jacket, private enough to hold wild ideas and heretical doubts.

On the cover he labeled it simply "B." Notebook "A," begun about the same time, was devoted to geology. As a heading on the first page of "B" he wrote "*Zoonomia*," in genuflection to a book of that title published forty years earlier by his own grandfather, Erasmus Darwin. This first Erasmus was a well-known physician and popular poet, a colorful fellow of big appetites—libidinous, gouty, unconventional in his views—who fathered a pack of legitimate and illegitimate kids and wrote erotic verse about plants. The name Erasmus had been passed down to an uncle, and then to Charles's brother, while

Charles himself got a different legacy: the propensity for scientific speculation. *Zoonomia*, mainly a medical treatise, included a section in which old Erasmus had floated evolutionary ideas of his own, suggesting that "all warm-blooded animals have arisen from one living filament," and that the common lineage possessed a capacity "of continuing to improve by its own inherent activity," with those improvements transmissible from parents to offspring. Erasmus Darwin had never pressed this idea too far, nor clarified it, nor supported it with evidence, but now he served his grandson both as a family forerunner in transmutationist thinking and as a point of departure. Charles's version *would* be clear, persuasive, and eventually supported by evidence—or he wouldn't let it see print.

He wrote the notebook entries in telegraphic style without much concern for punctuation or grammar. There were insertions, cross-outs, abbreviations, bad spellings. He had trouble with "heredetary," and with "Scicily" (or was it "Siicily"?), an island less unique zoologically than the Galápagos, though harder to spell. He was brainstorming. The written words were just a record for his own memory. He started with big questions. "Why is life short." And, following Grandpa Erasmus: Why is sex so important? From there he went straight to a crucial insight—that the mixing bowl of sexual reproduction somehow allows creatures to vary. Offspring differ from their parents. Siblings differ from one another, unless they happen to be identical twins. Body patterns change slightly from generation to generation; so do aspects of intellect and instinct. One result: "adaptation." To what eventual effect? Put a single pair of cats or dogs onto an island, Darwin suggested, and let them breed there, slowly increasing in number despite

the pressure of enemies; then "who will dare say what result."
He dared, but only to himself: "According to this view animals,
on separate islands, ought to become different if kept long
enough." There was *something* about islands. Their simplicity
and isolation and anomalous faunas, like premises of a
thought experiment, helped clear the head.

For instance, take the Galápagos tortoises and mockingbirds, or the miniature foxes he'd seen in the Falklands. "Each
species changes," Darwin wrote. This alone was a bold statement, explicitly contradicting orthodox tenets, both in science
and in religion. Furthermore, changing species diverge continually from one another, he hazarded, producing the gaps
between genera and still broader categories of classification,
such as family, order, class: the diversity of life. On one notebook page he drew a rough diagram of a lineage, like a tree
trunk, splitting into limbs and branches. The end of each
branch he labeled with a letter, representing a species. Birds
and mammals, vertebrates and insects, even animals and
plants—they were all branches from a single primordial
trunk. His mind was flying. Then he wrote: "Heaven knows
whether this agrees with Nature: *Cuidado*." Not so fast,
Charles. Be careful.

What's remarkable in the "B" notebook, besides the private
evidence of his leap into evolutionary thinking, is the breadth
of facts, notions, sources, and themes that Darwin was already
pulling together, some of which would remain pillars of his
work and his arguments for decades to come. He seized on the
idea of adaptation. He saw that variation among offspring
made it possible. He grasped the significance of biogeography
(that is, geographical patterns of species distribution) and of
classification (how living things can be sorted into groups) as

evidence that transmutation and divergence of species have occurred. He called attention to rudimentary structures— limbs and organs that seem too small, too primitive to be useful, as though they haven't been fully formed, or have latterly fallen into disrepair. Such rudiments exist even among humans. Why do men have nipples? Darwin, restless seeker, wanted to know. Why do some species of beetle, especially on windy islands, possess good wings sealed uselessly beneath fused elytra (those shelly wing covers) that can never open? Why would a smart, busy God create something so dumb and wasteful?

Flightless beetles were puzzling enough, but he wondered also about flightless birds, with their nubby little wings—the ostrich, the penguins, the rheas he'd seen in Patagonia, the apteryx of New Zealand. "Apteryx," he wrote, "a good instance probably of rudimentary bones." He hadn't collected an apteryx during his New Zealand stop with the *Beagle*, hadn't even glimpsed one, and he didn't call it by its native Maori name, the kiwi. But he knew enough from his reading to mention it, a small piece of the great puzzle, whose place would be found later.

3

For two years he lived a strange double life, like a spy in the corridors of the British scientific establishment, which at that time was closely attuned to Anglican orthodoxy and grounded in the tradition of natural theology.

Biology hadn't yet emerged as a secular profession. Studying nature was considered a path toward piety. Many prominent natural historians such as Gilbert White, author of a mellifluous little book of observational lore titled *The Natural History of Selborne*, first published in 1789, were clergymen who preached on Sunday and watched birds or chased insects the rest of the week. A blacksmith's son named John Ray, after an education at Oxford (which was then an Anglican university, like Cambridge), had sounded the theme back in 1691 with his book *The Wisdom of God Manifested in the Works of the Creation*. William Paley reaffirmed it in 1802 with *Natural Theology*, subtitled *Evidences of the Existence and Attributes of the Deity, Collected from the Appearances of Nature*, a book Darwin had read for amusement during his time at Cambridge. Paley popularized the analogy of the divine watchmaker: If we find a watch lying on the ground, we infer that some intelligent craftsman has made it; if we find intricately designed and marvelously adapted animals and plants, we should likewise infer that some wise, powerful Creator has made them. A series of books called the *Bridgewater Treatises*, published during the 1830s, offered eight further statements, from highly respected researchers, of the same argument about God's wisdom, power, and direct role in creating the natural world piece by piece. One of those *Bridgewater* authors was William Whewell, a polymathic scholar and science philosopher whose influence spread in many directions, and who invented the very word "scientist." Whewell's treatise considered astronomy and physics "with Reference to Natural Theology."

Behind the Paleyite natural theology lay even deeper and older forms of conventional belief, such as essentialism, the notion that reality is undergirded by a finite number of "natural kinds," the essential patterns or archetypes of entities seen in the world. This one goes back to Plato. Following his influence, essentialists held that these natural kinds are discrete and immutable, and that physical objects are merely their inexact manifestations. Geometric shapes, for instance, were thought of as natural kinds—triangles being always three-sided, various in their minor characteristics (equilateral, isosceles, scalene) and forever distinct from rectangles or octagons. Inorganic elements were another example—iron being always iron, and lead always lead unless some alchemist found a magical way of turning it into gold. Animal and plant species were also considered natural kinds, rigidly demarcated and unchangeable, though individual dogs or chickens might be various *within* their hard-sided categories. The essential form of a species, according to this view, is more fundamental and durable than the individuals embodying it at a given time. That's what William Whewell meant when, in 1837, he wrote emphatically: "*Species have a real existence in nature*, and a transition from one to another does not exist." To believe otherwise was to reject an assumption that was interwoven with ecclesiastic teachings and ideas of civil order.

Whewell, whose interests and writings ran to geology, mineralogy, political economics, moral philosophy, and German literature as well as astronomy and biology, became one of the heavyweight intellectuals of his time. The comment about species is from his *History of the Inductive Sciences*, produced in a later intellectual generation and a more rigorous scientific spirit than Paley's *Natural Theology*. Other British scientists and philosophers contemporary with Whewell, such as John Herschel and John Stuart Mill, shared that lingering belief in natural kinds, buried beneath their disagreements about scientific method and logic. In France, the eminent comparative anatomist Georges Cuvier proposed a system of animal classification—sorting every species into one of four great *embranchements,* or groups—that also rested on essentialist assumptions. Finding order within the animal world meant, to Cuvier, reading the evidence in each species for its conformity with an underlying essence, not the clues suggesting change and divergence over time. A philosopher of science from our own era, David Hull, has traced this vein of essentialism in early nineteenth-century biological thinking. Hull concludes: "Seldom in the history of ideas has a scientific theory conflicted so openly with a metaphysical principle as did evolutionary theory with the doctrine of the immutability of species."

Darwin had read Herschel, as well as Paley, back at Cambridge. Whewell had been a professor of mineralogy there. Essentialism and natural theology were as thick in the air of Darwin's world as coal smoke and the scent of horse manure. True, those weren't the only contemporary perspectives on the natural world. The private medical schools of London and Edinburgh harbored wilder ideas during the 1830s, including some inchoate versions of evolutionary progressivism. But those institutions, which employed some professional anatomists who taught by dissecting human bodies, who lived on their salaries and not on inherited wealth, and who tended toward radical politics, were alien to Darwin, despite the family tradition of doctoring. He had tried medical training himself, in Edinburgh at age sixteen, following the footsteps of his brother (in the shadow of his father), and he hated it. After two years, bored by the lectures and appalled by the bloody operations done without anesthetic, Charles had scooted down to Cambridge for a drier, less gruesome education. While there, at Christ's College, he had drifted toward clerical ordination, not from any sense of vocation (he wasn't devout) or Church commitment (he'd descended from Unitarians on his mother's side and Darwin freethinkers like his father and old Erasmus) but by the least-worst logic that it would allow him to find some respectable niche as a parson-naturalist, after the model of Gilbert White. The *Beagle* trip had intervened. The ship carried him a long way from Christ's College but it eventually returned him to the same social context he had left behind, in which many of his scientific teachers, friends, and connections—John Henslow and Adam Sedgwick at Cambridge, Leonard Jenyns, the entomologist Frederick Hope, William Whewell himself—were Anglican clergymen. Even his scientific idol, Charles Lyell, had imbued *Principles of Geology* with an orthodox view of biological creation. During 1837 and '38, Darwin began steeling himself to shock and scandalize them all. His view of mutable species directly contradicted their essentialism and all the pious science-flavored theologizing that stood upon it. He poured his dark speculations into the transmutation notebooks while conducting himself outwardly as a clubbable young naturalist on the rise.

He cut back on his socializing, with apologies about being too busy, then added to his chores and his status by accepting a role as secretary to the Geological Society, under Whewell's presidency. He finished the manuscript of his *Beagle* journal (but it couldn't be published until FitzRoy's book was ready), and talked his way into another big publishing venture: a lavish compendium to be called *The Zoology of the Voyage of H.M.S. Beagle*. He would be the editor of this multi-volume *Zoology*, gathering contributions from his consulting experts, writing introductions and commentaries, commissioning expensive illustrations, to be financed with a grant from Her

Majesty's Treasury. He was now well embedded within the seamless matrix of government, Church, and gentlemanly science. Secretly, he continued talking to himself in the seditious notebooks.

When he'd filled notebook "B," he started a new one, in maroon leather and labeled "C," after which would come "D" and "E," each devoted to transmutation. He was reading widely in the literature of exploration and natural history, plus a diverse selection of books on animal and plant breeding, history, and philosophy of science; and he had begun putting cryptic questions to anyone who knew anything about the odd, targeted topics that interested him. He debriefed his father, a gusty source of lore on human mental attributes, and his father's gardener. He quizzed livestock breeders about variation and heredity among domestic species. There were so many unknowns to consider. How did inheritance work? What was the difference between species and varieties? What might be deduced from patterns of species distribution around the world? All the islands of Oceania have skinks with golden streaks, he noted. Wild pigs in the Falklands grow stiff, brick-red hair. The kingfisher of the Moluccas scarcely differs from European kingfishers, he wrote, except that its beak is longer and sharper. Were they separate kingfisher species, or just varieties? Cassowaries in New Guinea, tenrecs in Madagascar, geckos on St. Helena. There are no snakes on islands of the central Pacific, he wrote. Black rabbits, introduced onto the Falklands back in 1764, had yielded decades worth of variously colored offspring. Clues, clues, clues. What did they signify, how did they fit? The cuckoos of Java versus the cuckoos of Sumatra and the Philippines—species or varieties? He wanted every possible piece of relevant data, whatever the source. He went to the Regent's Park Zoo to see its newly acquired orangutan. He became a greedy amasser of seemingly unconnected facts. He busted his brain to connect them. It was an intense program of research and cogitation, all in hours stolen from his public commitments.

"The changes in species must be very slow," he figured, not nearly so fast as when domestic breeders select which animals they want to pair. Slow or not, there was a problem to address: When animals continue interbreeding freely, won't the adaptive differences get blurred away? If so, "all the change that has been accumulated cannot be transmitted." Maybe isolation somehow prevents that. Maybe sterility between different forms, like the sterility of hybrids in domestic breeding, allows the accumulated change to persist. By now he was making some cocky comments in the notebook about "my theory," although that was premature. His theory hadn't yet coalesced. He was still groping to see the scope of the phenomenon, let alone to find a mechanism that would explain it. "Study the wars of organic being," he advised himself. Imagine that mankind didn't exist and that monkeys, breeding, improving, eventually produced some sort of alternate intellectual being. Manlike but not Man, and transmuted from four-handed, arboreal animals. That was hard to grasp, sure, but maybe not so much harder than Lyell's idea about slow, incremental processes accounting for all the big effects in geology. Remember the apteryx, Darwin told himself. If New Zealand had been divided into many islands, would there now be many apteryx species?

Seventy-five pages into the "C" notebook, in spring 1838, Darwin's confidence swelled. Grappling with these questions, he admitted, was "a most laborious, & painful effort of the mind," the difficulties of which would never be solved without long meditation or by someone prejudiced against the whole notion. But once you grant that species "may pass into one another," then the "whole fabric totters & falls." Look around the world, Darwin coached himself. Study the gradation of intermediate forms. Study geographical distribution. Study the fossil record, and the geographical overlap between extinct creatures and similar living species. Consider all this evidence, he argued excitedly, and "the fabric falls!"

The fabric was natural theology. For him it *had* fallen. Behind where that drapery had hung, Darwin saw the reality of evolution. It wasn't just a matter of mockingbirds, rabbits, and skinks. It was the whole natural world. "But Man— wonderful Man," he wrote, trying out ideas on this most dangerous point, "is an exception." Then again, he added, man is clearly a mammal. He is not a deity. He possesses some of the same instincts and feelings as animals. Three lines below the first statement about man, Darwin negated it, concluding firmly that, no, "he is no exception." From that terrible insight, despite pressures and implications, Charles Darwin would never retreat.

4

Did it make him physically sick? Possibly. Darwin's work on the transmutation notebooks coincided with his early complaints about what became chronic bad health. The symptoms were mysterious—jumpy heart, nausea, vomiting, headaches, nervous excitement, inordinate flatulence—but real enough to make him miserable and to slow his work. Was he a hypochondriac? A neurasthenic? Had he been bitten and infected by some nasty disease-bearing bug during a *Beagle* stopover in Argentina? Many guesses have been made but nobody knows, to this day, what ailed him.

Just before the voyage, he had experienced some cardiac discomfort, possibly reflecting his high state of nervous anticipation. He seemed otherwise to be a healthy young man, and he stayed robust throughout most of those five years. He suffered seasickness, yes, and an occasional bout of bad stomach or fever, not unusual for a stranger to the tropics; but during shore time in South America he managed long stretches of adventuresome hiking and riding. Since his return he had gained sixteen pounds, a good sign that the food at the Athenaeum Club agreed with him. Then, in September 1837, he said in a letter to his old Cambridge mentor, John Henslow: "I have not been very well of late with an uncomfortable palpitation of the heart." His doctors had advised him to quit work and get a country vacation, he added, and he was taking their advice. "I feel I must have a little rest, else I shall break down." After a few weeks home in Shrewsbury, with his father and sisters, he reported again to Henslow that "anything which flurries me completely knocks me up afterwards and brings on a bad palpitation of the heart." Social gatherings flurried him. Intense conversations flurried him. Conflict, or the very idea of it, was highly flurrisome. Eight months later he repeated to his old friend W. D. Fox the same muted phrase he'd written Henslow: "I have not been very well of late. . . ." There was too much to do, too much to learn and consider. He couldn't afford being sick. But the workload of his *Beagle*-related chores, and the sense of dreadful mission involving transmutation, didn't help his stomach. To make life more complicated (though maybe he imagined it would simplify things), he started thinking about marriage.

Not about marrying somebody in particular—just about *marriage* as a state, a condition, a step in the progress of a man. Was it something he should do? The daughters of Leonard Horner don't seem to have tempted him; possibly they were too smart and lively. He didn't mention any favored candidate, but the question of marriage had risen big in his mind, partly because it was related to another that also seemed urgent: money. How would he, over the long term, pay his bills? He had to eat, he had to buy books. He thought that he wanted to travel again (more comfortably than on a crowded naval ship). His current allowance might cover all that, but not the costs of a wife and children, too. At this point, unaware of how wealthy and generous his scary father was, Darwin imagined that choosing marriage would mean resigning himself to the necessity of a salaried job. Doing what? He had never finished medical training and he was definitely unfit—given what he believed, what he didn't believe—to masquerade as a clergyman. He considered wangling a professorship at Cambridge, maybe in geology. Nerdy, systematic, prone to anxiety, he tried to work through his confusion about marriage and money as he was working through the idea of transmutation, by scribbling notes. Since he was parsimonious about paper, not just about time and energy, he did it on the blank sides of a letter from Leonard Horner. Maybe that was also his way of turning the page on the Horner daughters.

"If *not* marry," he wrote, topping one section; then he listed a scenario of advantages. European travel. He might go to America, do some geologizing in the United States or Mexico.

Or he'd get a better house in London, near Regent's Park, and work on the species question. He could keep a horse. Take summer tours. Make himself a specialist collector in some line of zoological specimens and study their affinities. It didn't sound bad. "If marry," he wrote—then another list, this one mainly disadvantages, as though he were talking himself out of it. "Feel duty to work for money." No summer tours, no getaways to the countryside, no large zoological collection, no books. Ugh. Could he tolerate this, living in London, in a little house full of children and the dreary food smells of poverty, "like a prisoner?" Cambridge might be better, if he could get a professorship. "My destiny will be Camb. Prof. or poor man," he thought. He was wrong. But his resignation to that pair of options suggests he wanted a wife pretty badly.

He needed to ventilate his brain. In late June 1838, he broke away from London and its pressures—his editorial work on the *Zoology*, his chores for the Geological Society, possibly also the secret notebooks, unless he tucked "C" into a pocket—and went off to Scotland for a bit of geological fieldwork. He visited Glen Roy, a valley in the Highlands famous for the strange unexplained terraces traversing its slopes. Vacationing or not, he was a keen observer and a restless theorizer. After eight days in Glen Roy, he had his own notion about the origin of the shelves, and once back in London, amid all the other work, he'd make time to write a Glen Roy paper. But on the way south he stopped again in Shrewsbury for a family visit.

Talking with his father, Darwin got some brusque, cheerful advice: Quit worrying about money, you'll have plenty, and get married before you're too old to enjoy the kids. Dr. Darwin himself had been forty-three when Charles was born. The good news about financial support helped to rearrange Darwin's thinking. He drew up another orderly page of marital pros and cons, and this time "Marry" headed the longer column on the left, "Not Marry" the shorter one on the right. Marriage would give him a constant companion and a friend in old age, who would be "better than a dog anyhow." It was intolerable to think of spending one's whole life on nothing but work. "Only picture to yourself a nice soft wife on a sofa with good fire, & books & music perhaps." The Horner girls didn't fit that picture. Turning the sheet over, he wrote: "It being proved necessary to Marry . . . When? Soon or Late." The other question he might have added was: Whom?

Before heading back to London he visited his cousins, the Wedgwoods (of the famous pottery business and the family fortune it had built), at their mansion in the next shire. It was the safest household he knew outside his own family home. Considering the gruffness of his father and the supportive amiability of his uncle, Josiah Wedgwood, maybe it was the safest household, period. And, oh, there were unmarried Wedgwood girls.

5

By then he had begun notebook "D," the third in his transmutation series. "Mine is a bold theory," Darwin wrote, meaning the big one about species, not the little one he'd just concocted about Glen Roy, "which attempts to explain, or asserts to be explicable every instinct in animals." Yes, it asserted that animal instincts and much more were "explicable," but it did *not* explain those phenomena; it merely noted the fact that species are interconnected by common ancestry. Darwin still hadn't proposed a mechanism for how transmutation occurs.

Noodling on, he recorded some facts about Muscovy ducks, white-headed Sussex cattle, glowworms, and again the apteryx. From the anatomist Richard Owen he gleaned that reptilian skeletal structure is very similar to avian skeletal structure, as evident in a young ostrich. But Owen wasn't inclined to make as much of the reptile-bird similarity as he did. "There must be some law," Darwin told the notebook, "that whatever organization an animal has, it tends to multiply & IMPROVE on it." But what was the law? He still didn't know.

Despite the time lost to his unexplained illness early in the summer, by autumn he was back in a groove. He finished the Glen Roy paper. He worked on another geology manuscript, related to those endlessly ongoing *Beagle* publications. He pondered transmutation and also, by testimony to another little diary, he "thought much upon religion." The entry is cryptic, but it's safe to assume he wasn't experiencing an accession of piety. Probably he was worrying over the conflict between religious dogma as filtered through natural theology and, on the other hand, the view of origins he now held. He cast about for facts, alternative perspectives, and authority, reading the journal of an expedition to eastern Australia, Edward Gibbon's autobiography, John Ray's *The Wisdom of God*, and three volumes of a biography of Walter Scott. He read books about birds, Mt. Aetna, physiognomy, epistemology, and Paraguay. And then, in September of that year, 1838, he picked up the sixth edition of Thomas Malthus's *Essay on the Principle of Population*.

He would have known something of Malthus already, by cultural osmosis, in the same way an educated person today is at least vaguely aware of Milton Friedman or Jean-Paul Sartre. His brother's favorite dinner partner, Harriet Martineau, was an ardent popularizer of Malthus's views. The *Essay on Population*, first published anonymously in 1798 and expanded in later editions, offered a political economist's dispassionate analysis that undergirded the Whig program of hard-nosed welfare reform. Easy charity was bad and pointless, according to Malthus. It only encouraged population increase among poor people, without generating any commensurate increase in the national supply of food. That caused prices to rise for everybody. Eliminate unquestioning relief, force the poor to compete as laborers or be locked into workhouses, educate them against the disadvantages of profligate reproduction, and the problem of mass poverty would be ameliorated, if not solved. This was Malthusian social logic. It entailed stern thinking and, with a little exaggeration or distortion, could seem even sterner. Darwin was a mild, generous soul and he may have found it, as described secondhand, too callous.

What he probably didn't know until he had Malthus's book in his hands was that it mentioned animal and plant populations as well as human ones. On the first page Malthus paraphrased Benjamin Franklin, of all people, to the effect that every species has a tendency to proliferate beyond its available resources, and that nothing limits the total number of individuals except "their crowding and interfering with each other's means of subsistence." Empty the planet of life, Franklin had posited, seed it anew with just one or two forms—fennel plants, say, or Englishmen—and within a relatively short time Earth will be overrun with nothing but Englishmen and fennel. The inherent rate of population growth is geometric—that is, any population can *multiply* itself by some factor, not just *add* to itself, with each generation. For humans, Malthus calculated, the inherent rate amounts to doubling a population every twenty-five years. For fennel, which sets hundreds of tiny fruits on each plant, the inherent rate of population growth is much higher. But the inherent rate is just a biological possibility; such extreme increases seldom happen. Under normal circumstances, on a teeming planet as opposed to an empty one, runaway population growth is prevented by what Malthus called "checks."

The ultimate check is starvation. For humans it results from the fact that, while population is increasing geometrically, ever-intensified efforts at land clearance and agricultural improvement only increase the food supply arithmetically. That is, the sequence 2, 4, 8, 16, 32, 64, 128 runs away from the sequence 2, 3, 4, 5, 6, 7, 8. But food supply directly limits population numbers only during famine. Another kind of check is voluntary: the decision to refrain from marrying, to marry late in life, or to practice birth control (of which Malthus, a wholesome parson of pre-Victorian views, didn't approve). Still other checks operate continually: overcrowding, unwholesome work, extreme poverty, bad care of children, endemic disease, epidemic, war, and anything else that might contribute to sterility, sexual abstinence, or early death. Generally speaking, Malthus wrote, you could boil them all down into "moral restraint, vice, and misery." Darwin read this and something went click. He was less interested in moral restraint and vice than in what "misery" might mean to a mockingbird, a tortoise, an ape, or a stalk of fennel.

He ruminated in his "D" notebook about "the warring of the species as inference from Malthus." The geometric population increase of animals, as of humans, is prevented by such Malthusian checks, he wrote. He imagined it all freshly. Take the birds of Europe. They are well known to naturalists and their populations are (or were in his time, anyway) relatively stable. Every year, each species suffers a steady rate of death from hawk predation, from cold, from other causes, roughly maintaining its net population level against the rate of increase from fledglings. Food supply remains limited, nesting space remains limited, but breeding, laying, and hatching continue to push against those limits. Everything is interconnected and uneasily balanced. If the hawks decrease in number, the bird populations they prey upon will be affected, somehow. With new clarity Darwin saw predation, competition, excess reproduction, death—and their consequences. "One may say there is a force like a hundred thousand wedges," he wrote, and that it's trying to "force every kind of adapted structure into the gaps in the oeconomy of Nature, or rather forming gaps by thrusting out weaker ones." The final result of all this wedging, Darwin added, "must be to sort out proper structure & adapt it to change."

In shorthand scrawl, he had his big idea. Years later he would articulate the details and call it "natural selection."

6

The wedges metaphor went into his notebook on September 28. And then an odd thing occurred, by way of visible aftermath to this momentous epiphany: nothing. Darwin held his cards close and kept a poker face to the world.

In private he continued the notebook ruminations, finishing "D" with a burst of comments on the "differences" (that is, variations) among offspring in consequence of sexual reproduction, and starting the next in his transmutation series, notebook "E," with some increasingly confident references to "my theory." His theory explained how those small differences can accumulate into adaptations particular to differing circumstances. His theory, he realized, would be quite a horse pill for other people to swallow. Trying to keep his thoughts compartmented, Darwin also began another notebook, labeled "N," which was devoted to "metaphysical enquiries" provoked by the scientific ideas he was considering. Does a dog have a conscience? Does a bee have a sense of communal responsibility? Is the human conscience just another form of inherited instinct, an adaptation for social behavior? Is the human mind just a function of the human body? Does the idea of God arise naturally in human minds from that instinctive conscience? Months earlier he had posed almost the same question about God and conscience—whether "love of the deity" might result simply from brain structure—and then had scolded himself delightedly: "oh you Materialist!" Now his materialism was getting deeper, firmer, less embarrassed. Still, he didn't feel ready to go public. There were already enough materialistic evolutionary radicals, he knew, amid the current political squabbling about Chartism, democratized medical education, changes to the Poor Law, and they weren't his kind of people.

It was the dizziest season of Darwin's life. He stopped writing letters to his friends and family. He kept busy at his *Beagle*-related tasks, guiding a volume of the *Zoology* into print and adding a preface to his own expedition *Journal*. He performed his duties as secretary of the Geological Society. His health was going bad, in some still unexplained way, and he needed rest. He trusted his most serious thoughts only to the notebooks. "Having proved mens & brutes bodies on one type," he wrote, it would be "almost superfluous to consider minds," adding, "yet I will not shirk difficulty." In early November, two themes dominated his notations: the importance of sex and the search for laws. Sexual reproduction (as distinct from vegetative reproduction or budding, whereby an individual microbe or plant reproduces itself exactly) entails the paradox of inherited variation—that is, slight differences between parents and offspring, as a result of the mixing of elements from two parents. Fundamental laws (as distinct from divine whim) govern the occurrence of variation and the transmutation of species. He wanted to illuminate those "laws of life." In the midst of this heightened sense of danger, excitement, and solitude, he did something uncharacteristically impulsive: He hopped on a train for Staffordshire, turned up at the home of his uncle Josiah Wedgwood, and asked his cousin Emma to marry him. It was a reckless leap toward safety.

His proposal surprised her. Emma was a sweet-spirited and pious thirty-year-old, on the brink of what in those days was considered spinsterhood. She and a hunchbacked elder sister were the last Wedgwood girls left in the house. She had known Charles nearly all her life, as the boy-cousin closest to her in age (though she was slightly older), and the families were cross-linked by multiple marriage connections. Charles's mother, who died when he was eight, had been the sister of Uncle Josiah, and just a year before this sudden offer to Emma, Charles's sister Caroline had married the eldest Wedgwood son, another Josiah. Even Charles's Wedgwood grandmother, his mother's mother, had been a Wedgwood by birth who married her cousin, another Wedgwood. First-cousin marriage was common in those days and those circles, though that's not to say people weren't aware that too much inbreeding could bring problems; otherwise they'd have been marrying their sisters and brothers. On the positive side, unions between cousins helped keep family fortunes together. So the pairing of Charles and Emma was an obvious one, in some ways. Would-be matchmakers in the Wedgwood family had probably pondered it more consciously than the principals ever had. Still, as these two cousins aged, it hadn't looked likely to happen. Charles had paid Emma some attention during his visit in July, though not with enough ardor to suggest that those few conversations were supposed to constitute courting. Now here he was out of nowhere—having done his private calculus of upsides and downsides and concluded he should marry *someone*—putting himself forward, humbly but abruptly, as her suitor.

The surprise went two ways. When she accepted him on the spot, he was startled. Then they both let the idea sink in. There was no whoop-de-doo echoing through the house that day. Emma felt "bewildered" rather than giddy, and Charles had a headache. Everyone else, both fathers included, clucked their conventional noises of approval. Charley and Emma, of course, how perfect.

It wasn't perfect. One imperfection was the discrepancy between Emma's fervent, Bible-based Christianity and Charles's recent free fall into disbelief. Charles himself didn't yet know how far that fall would drop him or where he would land. But his father had warned him, probably just months earlier, that a man with theological doubts should conceal them from his wife. Nothing to be gained for anyone, according to the hardheaded doctor, in giving a woman cause to worry about the salvation of her husband's soul. Things might go along well until one of them got sick, and then she would suffer miserably at the thought of eternal separation, making him miserable, too. Charles promptly ignored his father's advice (which may have been the most prescient thing, if not the wisest, that Dr. Darwin ever said to him), telling Emma at least something of his heterodox thinking. Most likely he didn't raise the topics of transmutation, monkey ancestors, the idea of the deity as an inherited instinct, or the conundrum of male nipples, but whatever degree of apostasy he confessed was enough that she called it "a painful void between us." Then she brightened up and thanked him for his candor, having reassured herself that "honest & conscientious doubts cannot be a sin."

Doubts? That was putting it politely. By this time he had a whole new framework of scientific and metaphysical convictions, not just doubts. But if she was willing to twine their fingers together across the void and ignore it, so was he. Nowhere on his clerkish lists of marital benefits had he posited that a wife should be a philosophical soul mate and a full intellectual peer. He told his friend Lyell, in a letter announcing the engagement, of his "most sincere love & hearty gratitude" toward Emma—gratitude for "accepting such a one" as him. This was probably a candid statement, more revealing than he wished: his love tepid but genuine, his gratefulness robust.

Back in London, he returned briefly to the "E" notebook before house-hunting and other domestic preparations swamped him. Near the end of November, with his usual bumpy punctuation, he wrote:

Three principles, will account for all
1. Grandchildren. like. grandfathers
2. Tendency to small change . . . especially with physical change
3. Great fertility in proportion to support of parents

Bare and elliptical, it was his first full outline of the three causal conditions for natural selection: (1) hereditary continuity across multiple generations; (2) incremental variations among offspring; and (3) the Malthusian factor of inherent population growth rate, producing so many unsupportable individuals. Put them together and you had an explanation of how species transmutation occurs.

So much for the notebook. In his personal diary, he wrote: "Wasted entirely the last week of November." Was he complaining, apologizing, or bragging jocularly about a newfound sense of lightness? In early December, Emma came to town and stayed two weeks with her brother and sister-in-law, during which she and Charles threw themselves into the merry fuss of setting up a household. Then she went back to Staffordshire. To the end of the year, he occupied himself with further house-hunting, a bit of reading, and being laid up intermittently by his mystery illness. Having decided the marriage question, he was now impatient for the wedding to occur. His letters to Emma were cheery. In one, at the end of a long day, he described himself complacently as "stupid & comfortable."

On January 29, 1839, they were married in a small church near the Wedgwood mansion. Charles's brother didn't come up from London for the event, and Emma's mother stayed home sick. Dr. Darwin and Uncle Josiah had arranged bounteous financial settlements, formalized in a document at the county records office: £10,000 from Dr. Darwin's deep pocket, £5,000 from the Wedgwood side, to be invested on behalf of the newlyweds at 4 percent annually. That meant Charles wouldn't need a job and they'd have servants in the house. They were young gentlefolk from affluent, provident families. The marriage ceremony was performed by Reverend Allen Wedgwood, a cousin to everybody. There was no reception, but not because the Wedgwoods couldn't afford a party. There was no honeymoon, but not because the couple didn't want to be alone.

Charles and Emma left Staffordshire that day. By way of matrimonial celebration, they shared sandwiches and a bottle of water on the train down to London. It was their chosen style. A quiet pair, disinclined toward ebullient display. And he had to get back to work.

The Kiwi's Egg

1842–1844

7

Think of it as a bird's egg, taking form slowly inside him. Ovulation had occurred. Fertilization had occurred. Now came growth, from the microscopic scale of a single ovum to . . . well, to whatever size it would reach before laying. Don't think of a hen's egg or a goose's, or even the hefty egg of an avian lummox like the ostrich. Since the ovum was natural selection and the bird was Charles Darwin, think of it as the egg of a kiwi.

The kiwis are long-beaked, globular, flightless birds, strange creatures with hairlike feathers who run around at night eating insects and worms. There are several species and subspecies, all embraced by the genus *Apteryx*, all endemic (that is, native there and nowhere else) to New Zealand. They belong to the ratite group, meaning that ostriches, rheas, emus, and cassowaries are their closest living relatives. The elephant birds of Madagascar and the moas of New Zealand, two sets of extinct giants, were part of that group, too. If these birds are all related and all flightless, you might ask, how did they arrive in such remote and disconnected places as South America (rheas), Australia (emus and cassowaries), New Guinea (more cassowaries), Madagascar, and New Zealand? The answer seems to be that they walked. The ratite lineage dates back to an era before the ancient southern supercontinent, now known as Gondwanaland, separated into its continental and island fragments. Traveling on foot, the ancestral ratites dispersed all across Gondwanaland and then, sometime later, the land fragments drifted apart. The giant birds rode away like penguins on an iceberg.

Compared with other ratites, the kiwis are small—no bigger than an overfed chicken. Taxonomists have disagreed about just how many species to recognize, under just what scientific names, but currently the consensus seems settled on four: the North Island brown kiwi (*Apteryx mantelli*); the tokoeka (*Apteryx australis*); the great spotted kiwi (*Apteryx haastii*); and the little spotted kiwi (*Apteryx owenii*). The last was named for Richard Owen, who in 1838 presented a multipart paper titled "On the Anatomy of the Apteryx" to the Zoological Society in London. Darwin, having heard at least some of Owen's paper, referred to it in his notebook "D." The most remarkable thing about the apteryx, Darwin thought, was its small respiratory system, suggesting that in the wild this must be a shy, patient, creepy little bird, with little inclination to exert itself much and therefore little need to breathe heavily. Owen had only a single specimen to examine, a male, and he was an anatomist, not a physiologist or a field naturalist; so he missed some kiwi traits just as peculiar as the reduced lung capacity. An extraordinarily acute sense of smell. A low body temperature, unusually cool for a bird. An odd mix of furtive and aggressive behavior. Darwin missed them, too, along with the single most notable fact about kiwi biology: These little birds lay humongous eggs.

A female brown kiwi weighs less than five pounds. Her egg weighs almost a pound—constituting, that is, about 20 percent of her total weight. Among some kiwis, the egg-to-body weight ratio reportedly reaches 25 percent. A female ostrich, by contrast, lays an egg weighing less than 2 percent as much as herself. Certain other avian species—hummingbirds, for instance—lay more ambitious single-egg packages than ostriches, but few if any match kiwis. Relative to her body size, on a standard with other birds, the brown kiwi's egg is about six times as big as it should be. It contains also a disproportionate allotment of yolk, on which the chick will survive just after hatching. This egg takes twenty-four days to develop and, once it has, fills the female like a darning egg fills a sock. Having gorged herself for three weeks to support the growth of such a large embryo, during the last two days she stops eating. There's no room in her abdomen for another cricket. "Sometimes the egg-bearing female will soak her belly in puddles of cold water," according to one source, "to relieve the inflammation and to rest the weight." She is painfully replete with motherhood.

An X-ray photo of a gravid female kiwi, taken fifteen hours before laying, shows this: a skull, with its long beak; a graceful S-shaped neck; an arched backbone; a pair of hunched-up femurs; and at the center of it all, a huge smooth ovoid—her egg—like the moon during a full solar eclipse. She herself is now just a corona. It seems impossible. How can she carry this thing? How can she deliver? Will it reward her efforts and discomforts, or rip her apart?

The size of the kiwi's egg raises interesting evolutionary

questions. For starters: Why is it so big? What are the adaptive advantages for kiwi females (and for males, who do much of the incubating) of such heavy investment in a single chick? How has the kiwi lineage changed over evolutionary time? Did the egg evolve toward largeness? Or did the bird itself evolve toward smallness—a shrinking ratite, descended from moa-sized ancestors—while the egg stayed as it was? If the bird shrank and the egg didn't, why not? Those questions could take us into a discussion of allometry (the study of growth rates and size disparities within organisms) and kiwi evolution, which might be amusing. But allometry isn't the point here.

The point is simply metaphor. Every time I see that X ray of the mama kiwi, I think: There's Darwin during the years of gestation.

8

By spring of 1842 he was a famous author, thanks to the surprising success of his *Journal* from the *Beagle* voyage (published in 1839), and a father of two, thanks to Emma. He had been elected a Fellow of the Royal Society, Britain's foremost scientific club. But he was still stuck in an ugly little row house in filthy, raucous London, and still slogging his way through the less glorious, more technical publishing chores that had followed from the five-year expedition. As for his transmutation theory, nothing. Nothing published, anyway. Nothing written except those disjointed notes and an occasional coy hint, in a letter to a friend, that he was working on the question of species and varieties. To his close colleague Lyell he had let drop his doubt that species have a divinely decreed beginning. In his *Journal* he had mentioned the Galápagos mockingbirds and finches, different species on different islands, but declined to speculate further on such a "curious subject." He wanted to tell people about his theory, and he didn't. It wasn't ready. He wasn't ready. He had finished with his transmutation notebooks, three years earlier, and let them sit. Among his more overt reasons for inaction on the "curious subject" were that he had been too hectically busy and too often sick.

The mysterious vomiting, headaches, and other knock-down symptoms continued to afflict him intermittently. He had resigned from his secretaryship of the Geological Society, citing bad health, a legitimate excuse but also one that allowed him to immerse himself more fully in his own work. Intellectual hobnobbing was fine for those with the stomach; he found it literally nauseating. He was over the loneliness he'd felt aboard the *Beagle*, satiated with the sort of chirpy socializing his brother enjoyed, and had begun the process of retreating from London scientific circles into a reclusive life of research, writing, and invalidism. His marriage to Emma, entered in such a pragmatic and passionless spirit, had started developing toward what it would eventually be: an extraordinarily close mutual devotion and an asymmetric dependency, with her serving as his chief nurse and protectress. Even before the later children (eight more of them) arrived, those roles were enough to keep Emma busy—and, it seems, satisfied. She didn't need to function as her husband's intellectual sounding board, or as his transcriber, or his copyeditor, to feel fully engaged with his life.

Besides, there was still that "painful void" between his thinking and her beliefs, which neither of them cared to accentuate. They knew that their disagreements about God, scripture, creation, and afterlife were wide and irresolvable. Three years earlier, not long after their marriage, Emma had written Charles an earnest letter, describing her struggle to come to terms with his science-driven impiety. She was ambivalent, she admitted. She wanted to feel that "while you are acting conscientiously and sincerely wishing and trying to learn the truth, you cannot be wrong." On the other hand, she couldn't always give herself that comfort. She worried that "the habit in scientific pursuits of believing nothing till it is proved" had blinded him to the importance of revelation. She wondered whether Charles hadn't been unduly influenced by his careless, doubting brother, Erasmus. She warned him gently of the danger to his immortal soul if, rejecting dogma, betting against orthodox views of spiritual reward and punishment, he was wrong. "Everything that concerns you concerns me," Emma wrote, "and I should be most unhappy if I thought we did not belong to each other for ever." He didn't want her to be perpetually unhappy, not in this life, let alone any other. So he preferred to let the matter drop—at least until he published his theory, whenever that might be.

But he never forgot her letter. He saved it among his private papers, in fact, and occasionally pulled it out to reread.

For the present, he needed to focus himself on immediate tasks and conserve strength. His volume on coral reefs would be published any month. He was offering an ingenious, well-supported explanation for how they are formed. Next he would do a book on volcanic islands. Both of those had been added to his original ambitious plan for the *Zoology of the Beagle* series. Eventually he would write three complete volumes of geological observations from the voyage, plus editing five volumes on the zoology. All this took time; years. Where did the days go? In his diary he tried to keep track. The coral reef book alone, Darwin figured, cost him twenty months of effort. That was spread across four years during which he had worked also on the *Zoology*, the Glen Roy paper, some other geological projects, and (marginally) transmutation, losing the rest of his workdays to illness. Being a husband, a father, and a householder also took time, notwithstanding the help supplied by a butler, a cook, a nurse for the children, and other servants, as well as Emma's own indulgence of his detached, contemplative habits. In May, he and Emma bustled their gang off to Staffordshire for a vacation at her family home. After a month there, he shuttled over to stay with his father and sisters in Shrewsbury, leaving Emma and the kids behind.

He had left his notebooks behind too, in London, but that didn't stop him thinking. The holiday from other work became a chance to put something on paper about transmutation. During those summer weeks of 1842, amid Emma's family and then his own, he found enough quiet hours to write a dense précis of his ideas and of the evidence and arguments he'd collected to support them. He worked in pencil. This "sketch," as he called it, came to thirty-five pages. Unlike the notebooks it was carefully structured, moving from topic to topic in a way meant to build his case clearly and cogently; but like the notebook entries it was elliptical, with phrases and sentences suggesting much more (at least to him) than they actually said. It was an outline, an extensive one, of the book he intended to write.

He began with the topic of variation among domestic animals, noting the obvious point that individuals differ slightly from one another in size, weight, color, and other ways.

Because some of those differences are heritable, human breeders have been able to perpetuate and even amplify desirable traits by carefully selecting which animals to pair. With enough selection over long stretches of time, breeders even produced new races—speedy horses versus dray horses and tallow cows versus beef cows, for instance. This was the setup for Darwin's crucial analogy.

From variation among domestics he moved to variation among wild creatures, and to what he called here "the natural means of selection." Variation in the wild might not be as common or as extreme as variation among domestics (so he thought), but under certain circumstances it did occur. What caused it? He didn't know—and, for the present, that didn't matter. Some of those variations, like the ones among domestic animals, were heritable. Given the inherent rates of population increase and the enormous excess of insupportable offspring, to which Malthus had awakened him, wild creatures would be subjected to an automatic sort of culling, based on their capacities to compete for survival and for mating opportunities. By now he had hit upon not just his analogy, with domestic breeding, but his chosen term: "natural selection." The net result over thousands of generations, he wrote mutedly, would be to "alter forms."

He had described a physical mechanism (or at least, part of it) by which new species *could be* produced. But was there empirical evidence that they *had been* produced, one from another, through any such pageant of organic change? Yes, and in the second half of his draft he sketched that evidence, category by category: the fossil record, geographical distribution, systematic classification of species based on morphological resemblances, rudimentary organs (such as the wings of the apteryx), all of which tended to affirm the idea of transmutation and to belie special creation. Then he wrote a conclusion, highlighting as a sample case three species of Asian rhinoceros—those from Java, Sumatra, and India—and noting that a creationist would believe all three originated, with their "deceptive appearance" of close kinship, from separate acts of divine will. As for himself, Darwin wrote, he could just as well believe that the planets revolve in their orbits "not from one law of gravity but from distinct volition of Creator." If all species are handmade by God, then a person might also assume that Mars and Jupiter fly around because He's playing them like yo-yos. That's unlikely. Maybe even blasphemous. Wasn't the deity, if any existed, too sublimely transcendent for what we'd now call micromanagement? Darwin was suggesting an idea even larger than natural selection: that the universe is governed by laws, not by divine whim, and that the transmutation of species by natural selection is merely one of those laws.

He finished the rough sketch with a burst of eloquence. It was oddly consoling, Darwin noted, that from the hard Malthusian struggle involving "death, famine, rapine, and the concealed war of nature" had come a great good, the creation of the higher animals. "There is a simple grandeur," he wrote,

in the view of life with its powers of growth, assimilation and reproduction, being originally breathed into matter under one or a few forms, and that whilst this our planet has gone circling on according to fixed laws, and land and water, in a cycle of change, have gone on replacing each other, that from so simple an origin, through the process of gradual selection of infinitesimal changes, endless forms most beautiful and most wonderful have been evolved.

He had made a big move toward putting his thoughts forward. But it was only a private memo to himself. And even in private he had fudged on one thing: no mention of the origins of man.

9

Late that summer London was a mess, more so than usual, with police and Guards units on alert against possible rioting by Chartist demonstrators. A radical editor was tried and convicted of publishing "impious doctrines" such as atheism and socialism, as flavored in his news rag with a vague, political version of transmutationism. Across the country, half a million workers had gone out on a general strike for the Chartist demands, and military units were moving north to restore order in mill cities like Manchester. Troops in London faced off, with fixed bayonets, against hollering protesters not far from where the Darwins lived. It seemed the right time to do what Charles and Emma had been contemplating for a year: buy a home in the country and get away.

After some careful house-hunting, they chose a place in a somnolent little village called Down, in Kent, sixteen miles southeast of central London. Sixteen miles then meant two hours by horse and carriage, distance enough to give them tranquility but allow Darwin to commute back on special occasions for scientific business. The property itself, known as Down House, had once served as the village parsonage; lately it stood vacant, musty and unsold. It was a big house with multiple bedrooms, a fixer-upper at a bargain price, and it came with 18 acres of land. Helped by a loan from Darwin's father, they grabbed it. By late September they were in residence, not knowing that this would be their sole home and treasured refuge for the rest of their lives. Darwin himself may have hoped exactly that. The *Beagle* voyage had sated his appetite for travel and he felt ready to be a homebody. His wife was less enthusiastic about this drab house and the flat Kentish landscape around it, neither of which were impressive to a young woman raised on a fancy Staffordshire estate; but she figured she could adjust to it. The first major event in the new location was cheery, when Emma gave birth to a girl, their second daughter, christened Mary Eleanor. The next came like a bad omen, three weeks later, when the baby died. They buried Mary Eleanor in the Down churchyard. Now, in a grim way, they were rooted here.

The village Down later became Downe, with a spelling change meant to make it more distinctive. Darwin transmogrified himself, too, though not in order to stand out. On the contrary, he settled into village life as though it were a witness protection program. Assuming the trappings of a minor country squire, he planted flowers, bought a few milk cows, started an orchard, hired a handyman, took a seat on the parish council, established his private workspace in a study filled with books and files, and commissioned renovations for the rest of the house. Outside one window he attached an inconspicuous mirror, angled so that he could see people coming up the drive before they saw him. Visitors were hell on his weak gut, plus they cost him time that he needed for work. He didn't want company, except in very limited doses and on his own controlled terms. Lively chat made him excited and excitement made him sick. His study included a little lavatory nook behind a curtain, where he could vomit. From now on, most of his scientific conversations would be conducted through the mail.

He'd always been an exceptionally good letter writer, in a letter-writing age. Telephones didn't yet exist, after all, and any literate Victorian necessarily scrawled lots of missives to family, colleagues, and friends. Having a dinner party? Invitations went by note. Gossip and professional chat were largely epistolary, even among those who lived not far apart. After the move to Down House, Darwin took that a step further. Self-sequestered inside both his home and his sense of frail health, he became very dependent on written correspondence and very disciplined in his use of it. He wrote letters for friendship, letters for business, letters for love (to his "dear old Titty" or his "dear Mammy," as he variously called Emma, when they were apart), letters for good deeds and scientific politicking, letters asking parental advice and (later, with his sons away) giving it, letters for the sheer joy of prattle, and most of all, letters seeking scientific information. He peppered friends, acquaintances, and strangers with questions, requests for data, little assignments of experimentation that they might perform for him if, ahem, it wasn't too much trouble. He was unctuous and apologetic, but demanding.

What color are the horses of Jamaica? he wrote to a bureaucrat who'd once owned an estate there. Can you help me with an identification of certain rock specimens? he wrote to a professor of mineralogy at Cambridge. Your ideas about classification are airy and confused, he told George Waterhouse, the Zoological Society curator who had agreed to work on his

Beagle mammals, and who subscribed to a system of arranging similar species in neat circles, as though the deity had strung each genus into a closed loop like a pearl necklace. Darwin's language to Waterhouse was cordial, but his position was firm. The problem with those circles, he explained, was that they meant nothing and went nowhere. Darwin's own view, a controversial one he'd been keeping discreetly unspoken, was that "classification consists in grouping beings according to their actual *relationship*, ie their consanguinity, or descent from common stocks." That is, the underlying principle is transmutation. Saying so to Waterhouse, who wasn't among his closest friends, reflected Darwin's impatience to share his secret with *someone*. And then, in late 1843, he exchanged his first letters with a bright botanist named Joseph Dalton Hooker, just returned from serving as assistant surgeon and naturalist aboard a British ship, the *Erebus*, on its expedition to Antarctica.

Darwin had met Hooker passingly back in 1839, before the *Erebus* sailed, and knew something of this younger man from mutual friends. Hooker knew more about Darwin, having read his *Journal*, carried it on shipboard for four years, and idolized the scientific traveler who wrote it. Now they connected more personally—though only by mail—in regard to Darwin's old plant specimens from the *Beagle*, which had never been properly studied. Despite the chores Hooker faced with his own haul of specimens, he agreed to do it. Darwin asked him to pay special attention to the Galápagos plants, which might bear comparison with the peculiar species of St. Helena, another remote island. That suggestion triggered an outpouring from Hooker about native plants he'd seen on various islands while the *Erebus* circled the southern oceans, stopping in New Zealand, Tasmania, the Falklands, Hermite Island off Tierra del Fuego, Auckland Island, Campbell Island, Kerguelen, South Shetland, Ascension, and St. Helena itself. Hermite Island, for instance, was rich in mosses. Ascension held eight species of fern, only two of which occurred also on St. Helena, the next island over. Tasmania and New Zealand were unusual in ways of their own. Hooker went on for several pages, his overall message clear: If it's insular floras you want to talk about, sir, I can oblige you with enthusiasm and data.

Darwin claimed to be ignorant of botany and waited for Hooker to see the *Beagle* stuff. Hooker wrote again soon, expressing particular fascination with Darwin's Galápagos plants, clipped and pressed almost a decade earlier. From having read Darwin's comments in the *Journal*, he'd been well prepared to see floral differences among the respective islands, and with the specimens in his hands, that expectation was confirmed. The island-by-island diversity was, in his words, "a most strange fact." So strange, he volunteered, that it "quite overturns all our preconceived notions of species radiating from a centre." He meant a center of special creation, presumably on a mainland somewhere. No, the Galápagos plants were flat-out puzzling. Their botanical geography didn't jibe with the received wisdom of natural theology, and Hooker was willing to say so.

Darwin perked to that signal. He barely knew Hooker, but suddenly he felt a dawning hope that he had encountered a kindred mind. Hooker was smart and well trained, a conscientious observer; he came from a respectable scientific family (his father was director of the Royal Botanic Gardens at Kew), and had seen as much of the world as Darwin. Yet he was young (only twenty-six) and open to the possibility of junking orthodox tenets if empirical data so dictated. Darwin virtually grabbed him by the lapels. Early in 1844 he wrote again, asking Hooker's help with "one little fact" about endemic island plants. Then he ended his letter with an impetuous blurt of candor.

This is a famous moment. It appears in all nine of the Darwin biographies now piled on my desk, plus countless other studies, and it can't be omitted merely on grounds that the hands of previous writers and scholars have worn it smooth. The letter was undated, but the postmark said January 11, 1844. Darwin confided to Hooker that, besides his interest in southern lands, "I have been now ever since my return engaged in a very presumptuous work," a work that most people would call downright foolish. He'd been pondering the odd patterns of plant and animal distribution that he had seen in the Galápagos and elsewhere; he'd been reading up on domestic breeding; he'd been collecting every bit of data that seemed relevant to the question of whether species are changeless entities. "At last gleams of light have come," Darwin wrote, "& I am almost convinced (quite contrary to opinion I started with) that species are not (it is like confessing a murder) immutable."

This was a daring admission, cast in sheepish understatement, and contradicting one of the fundamental tenets of British natural theology. Truth be told, he was more than "almost" convinced.

Less famous is the disclaimer he added immediately: "Heaven forfend me from Lamarck nonsense of a 'tendency to progression' 'adaptations from the slow willing of animals' &c." He was trying to distance himself from the discredited ideas of one particular precursor, Jean-Baptiste Lamarck. Darwin knew well that his theory, besides being unsavory, might too easily be confused with other unsavory transmutationist notions that even he considered worthless.

10

Historians of biology have found intimations of evolutionary thinking in the works of philosophers and scientists long before Darwin. Books have been written tracking the concept back as far as Aristotle. Some of those early statements referred not to biological transmutation but to loosely parallel matters of cosmology and geology, such as the progressive physical history (from stardust to molten gob to rocky sphere) of planet Earth. Some involved the question of life's ultimate origin. Some were more closely related to evolution in the modern sense—that is, assertions about the diversity and classification of species, about continuity within that diversity, or about the tricky issue of just what a species is.

During the eighteenth century in France, for example, Maupertuis tossed forth the idea that vast numbers of living things come into existence by spontaneous generation, of which only a small fraction prove to be well organized enough for survival. Buffon articulated the hypothesis that apes, humans, horses, asses, and all other animals might be related by common descent—and then, having made it sound half-plausible, he backed away from that hypothesis. Diderot published dreamy speculations about living matter, generated in simple form but with a mystical sort of awareness, somehow assembling itself into complex creatures. In Germany, an anthropologist named J. F. Blumenbach studied skulls and suggested that the various races of humans had diversified from common stock in response to local conditions. In England, near the end of the century, Erasmus Darwin published his *Zoonomia*, with its casual suggestion about "one living filament" from which every sort of warm-blooded animal had arisen. All these bold musings added to an atmosphere of alternate possibility, offering at least some encouragement to anyone inclined toward challenging the rigidly scripture-based dogmas of creation. The likelihood of such challenges also increased with the arrival of new data: specimens and accounts of strange, unexpected species in remote places, sent back from the journeys of exploration and imperial conquest; volumes of biogeographical information, showing that new species and familiar ones are distributed around the planet in curious patterns; more and more fossils unearthed, revealing episodes of extinction and succession over time; the discovery, through microscope lenses, of tiny creatures swimming in every drop of pond water and saliva; the intricate adaptations seen in so many species; and the accumulating evidence of variation *within* species as well as differences among species. Despite all the restless speculation and all the new data, though, no one had proposed a comprehensive theory of evolution until, at the turn of the century, Lamarck did.

His full name was Jean-Baptiste-Pierre-Antoine de Monet, Chevalier de Lamarck, reflecting a family lineage that gave him trappings of nobility but no inheritance. At age seventeen he dropped out of a Jesuit seminary and joined the army. After a taste of war and a try at medicine, in Paris, he made himself a botanist, publishing an excellent three-volume flora of France. The book was well received but didn't solve Lamarck's problem of making a living, so he served two years as tutor and traveling companion to Buffon's son. Then he got himself hired as a botanical assistant, for a measly salary, at the Jardin des Plantes (which was later subsumed within the Musée National d'Histoire Naturelle). Lamarck's next metamorphosis, the most drastic one, didn't happen quickly. After twenty-five years as a botanist, he shifted to zoology, taking a museum position as professor of invertebrate animals and managing, throughout the Terror phase of the French Revolution, to keep his head down and away from the guillotine. His job was to lecture on insects, worms, and microscopic animals. Several years later the museum's mollusk collection fell into Lamarck's care when the malacologist, a friend of his, died. Studying that material, an assortment of fossils and recent shells, he saw evidence of variation within species and of sequential similarities among species found adjacent to one another in the column of time.

Abruptly, for whatever reasons, at the age of about fifty-five, Lamarck lost his belief in the immutability of species. Soon afterward, in May 1800, he gave his first lecture with an evolutionary slant. He presented his full theory nine years later in *Philosophie zoologique*, the book from which it's mainly known. A refined version appeared still later, in the introduction to his seven-volume natural history of invertebrates. Lamarck outlived four wives, went blind, survived to the age of eighty-five under the care of an unmarried daughter, struggled financially the whole way, and died in 1829, at which point he was more admired by radical British evolutionists (such as those teaching anatomy to medical students in Edinburgh and London) than by his colleagues in France. He was buried cheaply in an unmarked grave, like Mozart.

Most people, if they know anything about Jean-Baptiste

Lamarck, associate him with a single idea: the inheritance of acquired characteristics. There was more, as Darwin's groaning dismissal of what he considered Lamarckian nonsense ("tendency to progression" "adaptations from the slow willing of animals") in the letter to Hooker reflects. Lamarck argued that two factors account for evolution. One is, as Darwin noted, an inherent tendency in living creatures to progress from simple forms toward complexity. This tendency is conferred on them, Lamarck thought, by "the supreme author of all things." The simple forms originate by spontaneous generation. The increasing complexity comes as certain "subtle fluids" somehow open new channels through body tissue to create new and more intricate organs. Lamarck didn't explain why the progressive tendency exists, or just how those precious bodily fluids do their magic. He treated this factor as a given. It yielded separate lineages, progressing independently toward more complex species—but not to a branching tree of life. That's an important distinction to keep in mind: Lamarck never proposed that all creatures are descended from common ancestry. The right image for his theory would be prairie grass, with short stalks and long stalks rising parallel from the ground, not a bush or a tree with divergent branches, like the drawing in Darwin's "B" notebook.

Lamarck's second factor, which is more materialistic than his supposition of a God-given tendency to progress, encompasses four elements. First, animals face certain pressures from the external conditions (that is, the environment) within which they live. Second, when external conditions change, animals have new needs (*besoins*); they respond to those needs by increased use of certain organs or capacities, or by neglecting to use those they've been using. Third, increased use tends to enlarge or strengthen an organ or capacity; disuse tends to make it atrophy. Fourth, all such acquired changes are heritable. So here's the familiar idea, accurately associated with Lamarck but incompletely representing his theory: Offspring inherit the traits that their parents have acquired. The young giraffe is born with a long neck because its mother and father stretched to reach high leaves. The blacksmith's daughter is gifted with big muscles because her dad developed his over the anvil. Kiwis have useless little winglets because kiwi ancestors neglected to fly.

Two factors accounting for evolution, four elements within the second factor—and as though that's not enough, here's another ingredient of the theoretical stew: Lamarck's *sentiment intérieur*. At one point in the *Philosophie zoologique*, he posited this powerful but obscure *sentiment* (a sort of "feeling of existence," he explained, without adding much clarity) in higher animals, supposedly driving their subtle fluids and impelling their bodies toward those uses that produce new strengths and capacities. Maybe *sentiment intérieur* was just another name for what's now called consciousness. Or maybe he meant something more. Given the wooziness of the ideas and the losses in translation, it's not surprising that Lamarck has been often misconstrued. One misconstrual was that he claimed animals have an inherent power to enlarge organs or capacities in response to their *wants* (a misreading of the French *besoins*). A giraffe wants a longer neck so it can browse on acacias—and desire plus effort makes it so. That seems to have been Darwin's impression when he ridiculed Lamarck for suggesting that adaptations derive from "the slow willing of animals."

Darwin had gotten his first sniff of Lamarckism back in

Edinburgh, as a teenager, during the period when he was discovering that natural history engaged him much more than
the grisly and boring demands of medical training. He read
old Erasmus's *Zoonomia* and, in an uncritical way, admired it.
(He wasn't yet the tough judge of theory and supporting data
that he would be later, and it was nice knowing that his own
grandfather had written a notorious book.) He also read
Lamarck's technical work on invertebrate classification and,
more important, heard talk about Lamarckian evolutionism
from a dazzling young instructor who had befriended him,
Robert Grant.

Grant was crusty and formal on the outside but daringly
unconventional in his thinking; a prickly and complicated
man. Trained as a doctor, he taught invertebrate anatomy in
Edinburgh and spent his free hours doing research on marine
animals, especially sponges, or taking part in small scientific
clubs such as the Plinian Society. He had a habit of making
himself the mentor of select students, and in 1827 he picked
Darwin. Probably the fact that this gawky youth was the
grandson of Erasmus Darwin, whom Grant venerated as an
evolutionary pioneer, helped settle his attentions on Charles.
Together they took hikes to the seashore, waded in tide pools
collecting wormy and mossy creatures, dissected them with
the aid of the microscope at Grant's house, and eventually
shared a strong interest in one particular organism, the "sea-
mat" (but it was an animal, not an alga or a throw rug) known
as *Flustra foliacea*.

One day while they were walking, Grant launched into a
panegyric on Lamarck and his evolutionary theory, taking the
younger man by surprise. At that point, after all, Charles was
a dutiful middle-class kid from Shrewsbury who, despite the

surname, wasn't inclined toward radical ideas, especially those imported from France. "I listened in silent astonishment," Darwin recalled years later, "and as far as I can judge, without any effect on my mind." He hadn't bought transmutation as peddled wholesale by his own grandfather, and he didn't buy it now as retailed in Grant's riff on Lamarck. Another reason for his resistance may have been that he'd already seen a dark, mean side of Robert Grant, when the older man pirated some of Darwin's neophyte observations on the life history of that sea-mat, *Flustra foliacea*, and incorporated them into a paper. There was no acknowledgment to Charles Darwin, as data contributor or for anything else, in the published version. On the verge of his own first real contribution to science, Darwin had learned a hard lesson about credit and competition. He never forgot.

He met Lamarckism again during the *Beagle* voyage, when the second volume of Lyell's *Principles of Geology* reached him by mail in Montevideo. He had already read the first volume, which presented Lyell's critique of old-fashioned geological thinking, with its dependence on ancient catastrophes such as Noah's Flood. Lyell made the case for a new vision (adopted and modified from James Hutton's work, forty years earlier) of geological processes that were more continuous, more gradual, more uniform. That vision, in contrast to catastrophism, would become known as uniformitarianism. Lyell's point was that geological change tends to be slowly cumulative, not catastrophic, and caused by familiar forces that operate in the present as they did in the past. This seemed brilliantly persuasive to Darwin, and his own geologizing during the voyage was informed by it.

The second volume of Lyell's *Principles* was different.

Although it carried the same subtitle, announcing *An Attempt to Explain the Former Changes of the Earth's Surface, by Reference to Causes Now in Operation*, this one looked at flux and transition among the animal and plant kingdoms. How were fossils formed? How did peat grow? What went into the formation of coral reefs? Before tackling any of these questions, Lyell addressed a more controversial one: Do species themselves change? His first two chapters were devoted to Lamarck, giving a thorough exposition of the Frenchman's theory, noting that it "has met with some degree of favour from many naturalists," and then doggedly refuting it. No, species *don't* transmutate, Lyell decreed, not by action of Lamarck's factors nor by any others. Cats buried with Egyptian mummies look the same as our own cats, he argued. Feral cattle in America, living wild in an unfamiliar climate, eating unfamiliar foods, revert to exact likenesses of aboriginal European cattle. Yes, domestic breeding can yield novel strains of livestock, but those only constitute new varieties, never a new species. It just didn't seem compatible with Lyell's uniformitarian view—of large geological changes effected incrementally over long stretches of time—to recognize that species might change that way, too.

Darwin had read all this and agreed. Lamarck was nonsense. A dozen years later he still claimed to feel that way, with some coy stipulations: wrong answer, right question. But then how *does* transmutation occur? I think I know, he whispered to Hooker.

11

After confessing his transmutationist thinking and his murderer's sense of guilt, in that letter of January 11, 1844, Darwin waited for Hooker's response. None came. Two weeks passed. It's a safe guess that Darwin felt some nervous impatience. Had he scandalized the new pen pal? Had he kaboshed their friendship before it got going? Finally he sent another note, nudging Hooker to reply. And now Hooker did, with a long, cordial letter full of information on botanical geography and the cheerful news that he was studying Darwin's Galápagos plants. He chattered on about his pet subject, the native vegetation of remote islands, and mentioned a remarkable species of cabbage endemic to Kerguelen, in the far southern Indian Ocean. That Kerguelen cabbage, Hooker reckoned, was the weirdest cruciferous vegetable in the whole southern hemisphere. How did it get where it was? And why didn't it exist anywhere else? Pondering the cabbage, plus some other strange island creatures, led him to admit a heterodox opinion of his own: that there may have been a series of such unusual productions in isolated spots, "& also a gradual change of species." Whoa, change of species? That was a big concession. Sounding open-minded but sensible, Hooker added, "I shall be delighted to hear how you think that this change may have taken place, as no presently conceived opinions satisfy me on the subject." You're not crazy and you're not in danger of denouncement, not from me, he was telling Darwin, but . . . well, let's see what you've got.

Darwin went back to the sketchy outline he had tucked away two years before. He began writing again, articulating arguments and inserting evidence within the structure he had already arranged. This time he tried to produce something readable by others—and not just readable, persuasive. Again he started with the topic of variation among domesticated species, and how breeders select and amplify those small differences, the first leg of his cardinal analogy. Then he turned to species in the wild. He was still under the impression that wild populations vary little, except when unsettled by environmental changes. Never mind; a little was enough. Even tiny amounts of variation arising infrequently would allow natural selection, operating over great stretches of time, to produce new species of animal and plant.

Having explained his hypothetical mechanism—that is, the logic of how evolution *could* occur—he marched again through the categories of evidence showing that evolution, by some mechanism or another, *has* occurred. The pages piled up. Through late winter and spring of 1844, Darwin interrupted himself with a few minor distractions (seeing a short paper into print, commuting to London for Geological Society meetings, a family visit to Shrewsbury) but stayed focused and highly productive, as he could when his juices were hot and his health allowed. By early July he had finished a 189-page manuscript. This time he did what he hadn't done with the sketch: sent it off to a local schoolmaster to be copied in legible handwriting. The rationale for a clean copy was that it could be read by others. But by whom? Select friends like Hooker or Lyell? Typesetters at a publishing house? No . . . nobody, not for now. Instead he tucked it away in his office, along with a letter to his wife intended for reading "in case of my sudden death." The letter was his informal literary will.

Here's the draft of my species theory, the letter said. If the theory is correct, and if even one competent judge of the subject is converted, "it will be a considerable step in science." So please get this manuscript published, he told Emma, and he gave instructions on how he wanted that done. She should enlist an appropriate person to finish, improve, and edit the work. As enticement, she should offer that person £400, plus all Darwin's natural history books, plus any profits that might come from publication. She should also pass along to the editor all of Darwin's notes—his six years' accumulation of facts and quotes—written on scraps of paper and sorted by subject into eight or ten brown portfolios kept on shelves in his study. "Many of the scraps in the Portfolios contain mere rude suggestions & early views now useless," he explained, "& many of the facts will probably turn out as having no bearing on my theory." But he wanted his editor to sift through them. Some might be preciously relevant.

Who should this editor be? He mentioned a short list of scientific colleagues, including Charles Lyell, gentle old John Henslow in Cambridge, a brilliant paleontologist named Edward Forbes, and the congenial new acquaintance he'd been cultivating by letter, Joseph Hooker. The name of Robert Grant, Darwin's mentor from the Edinburgh days, now a hotfire radical teaching anatomy and preaching Lamarckism in London, didn't appear. Grant was a transmutationist, as Darwin well knew, but a transmutationist who embraced the wrong theory and belonged to the wrong political camp. Darwin wanted to modernize natural history, making it law-based and materialistic in its view of causes and effects; he did *not* want to foment class warfare. If none from his list agreed to take on such an onerous job, Darwin told Emma in the testamentary letter, she should please raise the offer to £500. And if that didn't suffice, he wrote, just publish the thing as it is.

For all he knew, the gut-heaving, head-blurring symptoms of his chronic illness might turn acute at any time, and he could be dead of some unknown ailment within a year. In fact, that may have been his subliminal wish. Dying now and publishing the theory posthumously would save him a lot of discomfort.

12

But he was getting closer, it seemed, to making the leap as a living author. He was getting bolder and more impatient. One day in July he made an unusual trip, by two-wheeled horse cart, all the way over to the Royal Botanic Gardens at Kew, southwest of London, to reacquaint himself with Joseph Hooker face-to-face.

Sickly and sedentary as Darwin generally felt, he wouldn't have gone if he hadn't badly wanted to cement his friendship with this young fellow. Hooker had several attractions. He was a rigorous botanist, well traveled, trained as a surgeon (not as a clergyman, like so many of Darwin's friends), and neither too scared nor too pious to contemplate transmutation. That's the person Darwin needed: a botanical geographer with the cold mind of a man who could cut human flesh. Through late summer and autumn he and Hooker continued trading letters, in which they discussed the distribution of species and why certain places—especially islands—contain an inordinate diversity of unique forms. Isolation is the crucial factor, Darwin suggested. The isolation of islands somehow leads to the "creation or production" (he was still waffling in his terminology) of new species. Darwin didn't explain what he had in mind, but he wanted Hooker's help in exploring this line of thought with botanical data.

Darwin also wrote to Leonard Jenyns, one of his parson-naturalist friends from student days, who styled himself after Gilbert White, keeping a diary of nature observations from the hedgerows and woodlands around his little parish. When they first met, back at Cambridge, Jenyns was a young fogey of thirty, recently established as vicar of a place called Swaffham Bulbeck and snugly ensconced in the tradition of natural theology. More recently he had edited a new edition of White's little classic, *The Natural History of Selborne*. Jenyns's next project would be a book of nature lore as collected by himself, including a natural history calendar, again in the manner of White. Swaffham Bulbeck was his own Selborne. Darwin flattered him about the importance of such localized, season-by-season observations, and then dangled a question that he hoped Jenyns might answer. How severely do struggle and early death limit population increase for any given species? For a species of bird, say, in the English countryside. He didn't mention Malthus, but of course it was Malthusian pressures and checks that Darwin had in mind.

There was more in the letter than just flattery and trolling for data. Jenyns had written to him first, a newsy note inviting news in return, so Darwin offered a glimpse of his present life and work at Downe. What with writing books about geology, he said, and looking after his garden and trees, and taking afternoon walks around the grounds with his brain in a fog, he hadn't lately done much field observation himself. No beetle collecting, like in the old days. He couldn't speak as an expert on local birds. Couldn't offer a single new fact about

English zoology. On the other hand, that wasn't to say he'd lost interest in flora and fauna. "I have continued steadily reading & collecting facts on variation of domestic animals & plants & on the question of what are species; I have a grand body of facts & I think I can draw some sound conclusions." Uh, but wait—did he really want to confide this to Reverend Jenyns? Evidently so. He was tired of *cuidado*. He was tired of keeping his secret. It poured out.

"The general conclusion at which I have slowly been driven from a directly opposite conviction," Darwin told Jenyns, "is that species are mutable & that allied species are co-descendants from common stocks." How about *that*, old pal? Evolution happens, and natural theology has missed the big story. I know this opens me to reproach, Darwin conceded, but I've been brought to it by honest and careful deliberation. "I shall not publish on this subject for several years," he added. His closing comment to Jenyns, sounding friendly, was almost a tease: Maybe your little local book will contribute something to my trove of supporting facts.

And then bad luck hit, in a form antic and unexpected as a rain of frogs. The same month as Darwin's letter to Jenyns, October 1844, a respectable London publisher released *Vestiges of the Natural History of Creation*, a volume of popularized science and theory-mongering that rampageously surveyed cosmology, geology, the origins of life, paleontology, and the transmutation of species, touching along the way such subjects as spontaneous generation, the rings of Saturn, the production of insects using electricity, the occurrence of measles in pigs, the origins of human races and languages, phrenology, six-fingered people, the germination of rye from planted oats, the birth of a platypus from a goose parent, the number of neck bones in a giraffe, plus many other interesting facts and astonishing factoids, all mixed and baked into a literary fruitcake by an author who wrote smooth, easy prose and who chose to remain anonymous. What curious reader could resist?

Thanks to its content and the mystery of its authorship, *Vestiges* became a hit. It raised eyebrows, stimulated thought, provoked annoyance, caused talk, and sold well. The scathing reviews it suffered from hardheaded scientists (including the great Cambridge geologist Adam Sedgwick, another of Darwin's early teachers) only added to its notoriety and spurred sales. An American edition appeared quickly, a German translation later. In Britain alone, *Vestiges* went into a second edition almost immediately, then a third, then seven more editions within a decade, totaling almost 21,000 copies. By the numbers of the day, that made it a blockbuster. It was read by middle-class gentlemen and ladies with no scientific or philosophical expertise, but also by Queen Victoria, John Stuart Mill, Abraham Lincoln, Arthur Schopenhauer, Ralph Waldo Emerson, Alfred Tennyson, Benjamin Disraeli, and Florence Nightingale. The fact that its author continued to guard his anonymity, not just during the early commercial success but throughout later editions, testified to the genuine riskiness of espousing transmutationism—even a godly version—if the chain of transmuted animals included humanity.

Vestiges wasn't atheistic. "It has pleased Providence to arrange that one species should give birth to another," the book said, "until the second highest gave birth to man, who is the very highest." Providence here was a law-making, non-intervening deity who established the physical universe and let it run. The author of *Vestiges*, a Scottish publisher named

Robert Chambers, likewise saw the wisdom of creating something and then staying out of sight.

Within two months of its publication, Hooker and Darwin had both read the book. Hooker breezily told Darwin that he'd found *Vestiges* delightful, not aware how that might make his friend cringe with the envy of a competitor. Of course he didn't swallow the book's conclusions, Hooker said, but the assemblage of material was impressive. As for the anonymous author, he seemed to be (Hooker didn't mean this as a compliment about wittiness) a "funny fellow."

Darwin saw nothing delightful or funny in any sense. He wrote drily from Downe that he'd been "somewhat less amused" by *Vestiges* than Hooker had. Okay, the organization was clever and the unidentified author could certainly write. But "his geology strikes me as bad," Darwin complained, "& his zoology far worse." This was a fair judgment, on scientific grounds, with a tincture of sour grapes. Darwin realized that "Mr. Vestiges" had just made his own position more difficult in ways that were both maddening and confusing. With its cockeyed pastiche of theory and its factual mistakes, the book gave credulous readers a misleading set of unsupported notions; it gave skeptical scientists another reason to dismiss transmutationism as bunk. Which was too bad for Darwin, and too bad again. Now the intellectual marketplace was glutted, the whole question was blurred, and the serious critics had their blood up.

Darwin may have hoped that the success of *Vestiges* could help open people's minds about transmutation; that it might prepare them to accept, in the long run, a *real* theory grounded in evidence and meticulous inductive thought. But that time frame—the long run—was speculative and remote.

For now, the moment for revealing his ideas seemed to have passed. He turned back to other projects. He had a third volume of *Beagle* geology to finish. He had a small *Beagle*-related task in zoological description, involving barnacles, to polish off. And he planned to revise his *Journal* for a new edition. Given a decent contract (unlike the one FitzRoy had arranged) with a different publisher, the book might actually earn him some money. If there was a best time for publishing his transmutation theory, this wasn't it.

Point of Attachment

1846–1851

13

If you view Darwin from a distance rather than close up, something peculiar happens now. He seems to stop. He seems to turn away. The idea of evolution by natural selection has been clear in his mind and in his notebooks since 1838. The extended essay of 1844 rests on a shelf in his office, unpublished. *The Origin of Species* will not see print until 1859. Meanwhile, as the years pass, he continues fathering children, pottering around the house, acting like a hypochondriac; he dissects barnacles through a microscope and raises pigeons in a coop. He publishes little papers in the *Gardeners' Chronicle* on subjects such as salt, bucket ropes for wells, fruit trees, a mouse-colored breed of ponies. Nothing on transmutation. He spends months at water-cure spas, allowing himself to be tortured with cold showers and wrapped in wet towels. It's the period of unexpected behavior that has been called "Darwin's Delay."

Scholars disagree about this period, and there's enough ambiguous evidence to nurture a whole range of possible explanations. Was he afraid to publish his theory because he knew it would outrage Victorian society? That's a lame generality, a first-draft cliché that ignores the diversity of Victorian society. Victoria herself had read *Vestiges*, after all, and though the author chose to preserve his anonymity, nobody was trying to find him and put him in jail. Robert Grant had been ranting about roughly the same stuff, in his lectures to medical students, for years. Was Darwin afraid to publish because of the political climate, in which the established Church and the government had reason to be wary of populist demagogues, Chartist mobs, maybe outright insurrection, as bolstered by Lamarckism and other subversive French ideas? It's true that Darwin had no love for extreme democratic ferment. He was a wealthy landowner himself, and a gentleman, a mildly progressive Whig with money and status to lose; he didn't want to sew any flag that political radicals might wave. Was he reluctant to publish because he came from the Oxbridge tradition of natural theology, within which many of his old friends and teachers were pious Anglican clergy? Was he just too polite to toss transmutation in their faces? Or was he hesitant because his wife, deeply pious, worried that his materialistic ideas would cost Charles his soul? Another alternative: Was he less anxious about transmutation per se than about the theory's logical extreme, *human* descent from a lineage of other animals? And then there's his undiagnosable bad health. Was he afflicted by some genuine disease or disability, with days of nauseated inertness on the sofa adding up into months of lost productivity? Or was the illness at least partly psychosomatic, his body's way of excreting the queasiness in his mind? Still another possibility: Maybe he proceeded slowly, deliberately, for good scientific reasons. Gathering data the whole time. Exploring complex implications of an idea that's not nearly so simple as it looks. Refining his arguments, running experimental tests, educating himself in unfamiliar areas of knowledge (taxonomy, embryology, animal husbandry) that would be crucial to making his case. Given the huge task of justifying a huge theory, was his rate of progress actually pretty respectable? Or, then again, was he just too busy for twenty-one years, diverted by all the various chores, projects, and human responsibilities that life brought him?

The answer to each of these questions, I think, is yes. The real uncertainty lies in how all the factors interacted—their relative importance, their intricate synergies—and that isn't likely to be settled by psychobiography or squinty textual analysis at a distance of a century and a half. Charles Darwin was a complicated man, courageous but shy, inspired but troubled, with a brilliant mind and a soft heart and a stomach that jiggled like a paint-mixing machine. If he were more unitary and transparent, he wouldn't be so interesting.

But a bit of tabulation and arithmetic, at this point, might help bring him into closer focus. In autumn of 1846, he was thirty-seven years old. During the decade since leaving the *Beagle* at Falmouth harbor, he had published three books, all dealing with the voyage: two geological treatises (one on coral reefs, one on volcanic islands) and his *Journal* from the *Beagle*. The *Journal*, a popular success, had lately gone into a second edition. His third geological volume (on South America) was in printer's proofs and would appear soon. He had also edited five volumes in his *Zoology of the Beagle* series and published about two dozen scientific papers. Most of the papers were short and slight, but the one on those strange terraces lining the slopes of Scotland's Glen Roy was long and ambitious, covering forty-two pages in the *Philosophical Transactions of the Royal Society*. In it he argued that the shelves were old sea beaches, formed when ocean levels had risen into the glen during ancient episodes of landscape subsidence; this agreed with his larger view, absorbed from Lyell, that rising and falling land levels play a big part in shaping geological features and placing fossil deposits. The Glen Roy paper, a major contribution to a prestigious journal, containing a bold theoretical assertion, was important to his reputation and self-image at the time, and important in a different way later when it proved embarrassingly wrong. In fact, you could add the Glen Roy embarrassment to the list of possible reasons why he delayed offering his theory of evolution.

The *Journal* was important too, and less ambivalently so, having made him a famous young scientific traveler back in 1839. It had originally been released as volume three of FitzRoy's set, under the dismissive title *Journal and Remarks*. Darwin was cast as a supportive voice to the main authors—FitzRoy himself and an earlier captain—of that *Narrative of the Surveying Voyages of H.M.S. Adventure and Beagle*. (After all, he'd begun the voyage in an unofficial capacity and succeeded to the naturalist's role almost accidentally.) But at publication time Darwin had stepped out toward the footlights and stolen the show because his volume, unlike the other two, was a good read, full of robust adventures amid exotic landscapes as told by an affable narrator. People liked it. Three months later, sensitive to demand, the publisher had reissued Darwin's volume alone, which must have put an additional crook in Robert FitzRoy's imperious snoot. Darwin's revised title was more expansive and confident, in the windy Victorian way: *Journal of Researches into the Geology and Natural*

History of the Countries Visited by H.M.S. Beagle. That edition sold well but, since Darwin was still bound by the contract FitzRoy had arranged, it never earned him a penny. Six years later he made a better deal on his own, signing over the copyright to a new publisher for £150, which in 1845 was real money. He did an energetic revision, cutting passages that seemed tedious, adding others that offered more flavor, inserting new results from the experts who had worked on his collections, reversing the order of *Geology* and *Natural History* in the title as a subtle reflection of the fact that geology was no longer his own primary interest.

His most notable changes to the *Journal* appeared in its Galápagos chapter. He added a drawing of four finches, showing the gross differences among their beaks, which John Gould had helped him appreciate. He wrote: "Seeing this gradation and diversity of structure in one small, intimately related group of birds, one might really fancy that from an original paucity of birds in this archipelago, one species had been taken and modified for different ends." In the earlier (1839) edition he had muttered a safe, vaguely theistic comment about how "the creative power" had been busy in the Galápagos. In the new (1845) text, he changed that to marveling at "the amount of creative force," a subtly different formulation, more quantitative than pious, and he admitted feeling "astonished" at the abundance of unique species inhabiting such a small archipelago, especially since the islands were formed by relatively recent volcanic action. "Hence, both in space and time," he wrote, "we seem to be brought somewhat near to that great fact—that mystery of mysteries—the first appearance of new beings on this earth." It was a teaser line. In mentioning "that mystery of mysteries,"

Darwin was echoing a phrase coined by the eminent science philosopher John Herschel; the mystery Herschel meant was "the replacement of extinct species by others," as evidenced in the fossil record but not easily explained by natural theology. Adopting Herschel's phrase gave Darwin a respected authority for viewing the origin of species as an unsettled matter, and it allowed him to hint his interest in solving that mystery. Then he shifted blandly to a discussion of Galápagos rodents.

Readers of the revised *Journal*, in 1845, were left to admire the finch drawing and wonder what the hell it meant. Maybe these islands did bring Darwin "somewhat near" the big question; but he wouldn't go nearer, not in print, for another thirteen years.

Despite his detachment from society and his devotion to science, Darwin liked earning money, and not just as an author. He kept a birdwatcher's eye on his investments, one of which was a 324-acre farm near a village called Beesby, in Lincolnshire, bought with inheritance money advanced by his father and eventually yielding profit as a rental property. Owning the farm made him a landlord, "a Lincolnshire squire," as he mockingly called himself. He also held shares in canals and, later, in railroads. At the start of their married life, he and Emma had received about £1,200 annually, mostly in interest on those gift trusts from their fathers. Of that amount, despite running a big household, they managed to save a little. Their income rose gradually for a decade and then, after Dr. Darwin's death in 1848, abruptly. The doctor's estate, split in unknown ways among the two brothers and their sisters, seems to have brought Charles a lump of about £45,000. That was a fortune. During the years immediately afterward, Charles and Emma's joint income went above

£3,700 annually, of which they managed to reinvest half. Their wealth continued adding up. Compared to the revenues from family legacies and savvy investments, his profits from book publications were small, though not too small to figure in his meticulous financial accounting. The *Journal*, after its unremunerative first edition, had brought him that modest but gratifying payment for the second. Suddenly he was not just a published author but a paid professional. He stuck with this new publisher, John Murray. *The Origin of Species*, fourteen years later, would be a financial success for both of them, as well as a towering scientific milestone. From just the first two editions of *The Origin* (released in late 1859 and early 1860), Darwin would make £616 13*s.* 4*d.* And that was only a start.

He wasn't miserly, just a bean-counter by habit. Details mattered. There are account books in Darwin's hand that show all his income and expenditures for forty-three years, from his marriage to his death, including such particulars as the £25 annual wages paid in 1842 to his butler, Parslow, and the 18 shillings he spent on snuff for himself in 1863. His outlay for shoes in 1863 also came to 18 shillings; shoes might be expensive but they lasted, even for a walking man, and snuff was his main vice. After five years at Down House, he put £58 into improving the garden and grounds. The beer bill for the household that year totaled £32. No tally exists of who drank how much.

In 1846, he had four surviving children—two boys, two girls—with another on the way. There would always be, until Emma was almost fifty, another on the way. She delivered ten children in all, of whom three would die young. Neither her rate of recurrent pregnancy nor the mortality among her offspring was unusual for the times. Darwin eventually became tormented, though, by concern over the health of his children (besides the three lost, several others were sickly) and a guilty sense that maybe they had inherited his bad constitution. He even entertained the dark notion that inbreeding—because he and Emma were cousins—might be part of the problem.

In the village, he was a pillar. He befriended the local curate, a young fellow who had just arrived during the mid-1840s, and played a helpful role in business affairs of the parish, though he stopped attending services, leaving that to Emma and the kids. Slightly later, he consented to be treasurer of the church's Coal and Clothing Club, and eventually also of a cooperative benefit society for the working folk, the Downe Friendly Society, founded at his suggestion. Expanding his own domain, he leased an additional strip of land along the back edge of the property, west of the big meadow, and planted it with birches, hornbeams, dogwoods, and other trees, plus a hedge of holly. Circled with a gravel path, it became known as "the Sandwalk," his daily route for cogitative strolls. The loop wasn't long, roughly a quarter mile, so sometimes he made a number of circuits, keeping track of his distance by kicking rocks onto the path like abacus beads each time he passed a certain point. He watched his children at play. He noticed birds' nests. He liked the tranquility and the balm of routine. He disliked provocation and upheaval. "My life goes on like Clockwork," he confided to FitzRoy, when they communicated for the first time in years, "and I am fixed on the spot where I shall end it."

FitzRoy was just back from New Zealand, having been sacked from his governorship there by the Colonial Office. Darwin's letter was written on October 1, 1846, one day short of ten years since he had jumped impatiently off the *Beagle*. If he was feeling a surge of nostalgia, plus some sympathy and lingering gratitude toward this man he had never found likable, he was also feeling something else: years passing quickly relative to his own pace of accomplishment. In his diary he noted the decade mark, and that he had just finished correcting the last page proofs of his *Geological Observations on South America*. The geology trilogy, by his reckoning, had cost him four and a half years. "How much time lost by illness!" he groused.

But during the healthy intervals he was a hard, steady worker—nowadays we'd call him a workaholic—grinding along without breaks for vacation or celebration between one project and another. He wasn't a man to uncork a champagne bottle and kick up his heels just because some book had been finished. On that same day he set the geology page proofs aside, October 1, 1846, Darwin turned to his single remaining container of preserved specimens from the *Beagle*. It held about a dozen barnacles of a very odd sort, minuscule creatures that drilled burrows into the shells of certain marine snails; he had collected them eleven years earlier in the Chonos archipelago off the coast of Chile. Now he meant to dissect these little beasties, get a grip on their identity, write a paper.

He started work under a happy delusion that it wouldn't take long. He didn't foresee being swallowed up by barnacle taxonomy for eight years.

14

From 1846 to 1854, he did almost nothing else but. At a low bench near one window of his study, seated on a revolving stool, he dissected barnacles through a microscope. He drew what he saw. He saved the parts of his dissected specimens on carefully sealed slides. He filed the slides in a drawer. He composed intricate technical descriptions, species by species. He read the barnacle literature, muddled and spotty as it was. He made decisions about how to classify the species he described, correcting bad choices of previous classifiers. Barnacles weren't easy. They came in two types, one type (the sessile barnacles) resembling limpets armored in boilerplate, the other type (the stalked barnacles) resembling mussels mounted on golf tees. Confusing things further, their larvae swam around like larval shrimp. Darwin wrote to barnacle experts and barnacle fanciers, cadging the loan of specimens that he would minutely dismantle before returning what was left. He commissioned a new style of dissecting scope from an instrument maker in London and paid £16, half a year's beer money, to get it built. His study must have smelled like a pub, from the evaporation of pickling alcohol off his specimens. His eyes were bleary at the end of a day's work. Emma delivered another daughter and three more sons during the barnacle period, superintending Down House and all its human activities while Charles labored fervidly. There's a story often told about the impression this eight-year effort made on his young children; the second son, George, visiting one day at the house of some playmates, asked them: "Where does your father do his barnacles?"

Barnacle taxonomy was an unplanned detour leading away from, then eventually back toward, transmutation. It started as a modest task, the describing of one species, and gradually became an obsession—something he wanted to do, something he had to do, something that wouldn't be done until it was done completely and right. But the detour wasn't random or accidental. The guiding impetus had shown itself earlier, within an exchange of letters between Darwin and Hooker in 1845, long before Darwin uncorked the Chonos bottle. They were discussing a certain book on the nature of species, by a French botanist named Frédéric Gérard. Evidently the work was slipshod. Hooker, always rigorous on botanical matters, told Darwin: "I am not inclined to take much for granted from any one who treats the subject in his way & who does not know what it is to be a specific Naturalist himself." Although Hooker meant no slight to his friend, only to Gérard, Darwin was defensive. Having established his credentials in geology but not in systematic biology—that is, he had never studied any single group of animals or plants with the plodding attention and descriptive purposes of a taxonomist—he took Hooker's comment as a personal criticism. Immediately he wrote back: "How painfully (to me) true is your remark that no one has hardly a right to examine the question of species who has not minutely described many." Thirteen months later, not wanting to seem a hare-brained theorist ungrounded in the details of how one species differs from another—like Gérard, or the author of *Vestiges*—Darwin began describing barnacles.

That first creature he worked on was puzzling in several ways. No bigger than a pinhead, it belonged among the limpetlike type, the sessile barnacles, he thought, but instead of cementing itself onto a rock and secreting a cone-shaped array of body armor, it found shelter by drilling into a snail shell. Darwin recognized that it represented an unknown genus, provisionally named it *Arthrobalanus,* and began mentioning it fondly as "Mr. Arthrobalanus" in his letters to

Hooker. After two weeks of dissecting he was charmed, feeling that he might spend another month on it and uncover some beautiful new structural surprise every day. He rigged a couple blocks of wood to support his wrists while he worked, and told Hooker how glad he was, following all the years of geological write-ups, to be using his eyes and fingers in this way again. After a month, having dug deeper, he was perplexed at the sexual peculiarities of Mr. Arthrobalanus. Most barnacles were known to be hermaphroditic, each individual carrying both male and female organs. This one, so far as Darwin could make out, had two penises and no egg sac. That was his first hint of what would emerge as an important finding from the whole study: Some barnacle species are hermaphroditic, some are separated into males and females, and some are frozen in complicated arrangements halfway between. The sex lives of the Cirripedia (the technical name for all barnacles), varying in progressive stages from hermaphroditism to distinct males and females, suggest a trail of transmutation.

In late November 1846, he sent a draft of his *Arthrobalanus* paper to the anatomist Richard Owen, asking for feedback and confessing that he had gotten so intrigued by barnacles that he was now dissecting a half dozen other genera. During the following spring he lost more weeks to bad health, this time including boils. He also interrupted his work with trips to London for the Geological Society and, the following June, to Oxford for the annual conference of the British Association for the Advancement of Science, one of the last of those big meetings he would attend. Most of his networking was now done by mail. He borrowed a sizable hoard of specimens from a rich private collector and made contact with several museum curators to get more. The consensus among experts was that barnacle classification was in disarray, and at least two of these men told Darwin that he was the guy to fix it. By the end of 1847 he had set himself to do a comprehensive monograph on barnacles, describing new species, revising earlier descriptions, bringing systematic order to the entire group.

He persuaded officials at the British Museum to send him their barnacle holdings on a long, trusting loan, and put out calls in every other imaginable direction. He even dropped a note to Sir James Clark Ross, the captain who had commanded Hooker's Antarctic voyage and was now preparing an Arctic expedition in search of Sir John Franklin, a fellow explorer, mortally stuck somewhere amid the frozen straits west of Baffin Island. While you're up there dodging icebergs and looking for Franklin, Darwin asked, would you please get me some northern barnacles? You'll be busy, of course, but it wouldn't take long to scrape a few off the rocks. Preserve them in spirits, he demanded sweetly, and make sure you don't damage their bases. Ross evidently ignored him.

The scientific confusion over barnacles included a disagreement about just where to place them within the animal kingdom. Are they mollusks? They seemed to be, given that they enclose themselves in shells (the adults do, anyway), live sedentary lives, and gargle seawater through their interior cavities. That misconception was corrected in 1830 by a researcher named J. Vaughan Thompson, who had noticed that the larval stages of barnacles, swimming freely, resemble crustaceans. By the time Darwin came along, it was agreed that the barnacles *are* crustaceans, more or less. The name Cirripedia (meaning "hairy-footed") reflects the fact that within each shelly exterior lurks a strange little being like a misshapen crawfish, with its head glued to the substrate and its wispy legs waving upward to grab food. Darwin's chosen task was to make sense of the Cirripedia, species by species, genus by genus, family by family, and to assign them collectively a rank and a place within the great phylum of joint-legged animals that in those days was called Articulata. Did the Cirripedia constitute a distinct class of articulates, separate from and parallel to the crustaceans, the insects, the arachnids? Or were they merely a minor division within one of the subclasses of Crustacea already known? Based on his close study of barnacle anatomy, making comparisons among species and matching larvae to adults, Darwin eventually called them a subclass unto themselves. Within the subclass he recognized two main families, the sessile barnacles and the stalked barnacles, plus several aberrant forms that fit into neither. One of the aberrant forms was Mr. Arthrobalanus, the starting point of all his barnacle travails.

Such decision making about categories and similarities represented the routine, necessary work of taxonomy, into which Darwin had plunged himself. The rationale behind this branch of biology is that the human mind craves order, and taxonomy (describing species, naming species, classifying them within a system of sets and subsets) is what gives comprehensible order to the dizzying multiplicity of living creatures. It's a very old game. Aristotle sorted animals into the "blooded" and the "bloodless" (insects were bloodless) and proceeded from there. He managed to distinguish whales from fishes, but also mistakenly separated them from mammals, and he included barnacles with mollusks. During the Middle Ages, plant identification and classification became important for medical purposes, and some experts published herbals (plant dictionaries) telling people what was what. Plants were simply listed, by name, in alphabetical order. But as the number of known plant species increased, herbalists found that alphabetization wasn't the most convenient or useful way to present their information. A man named Caspar Bauhin, offering notes on six thousand different plants in 1623, grouped species into genera according to their similarities of appearance or other physical traits; Joseph Tournefort, sixty years later, clarified the genus concept and placed his genera into classes. These relatively obscure contributors were followed by Carl Linnaeus, the famed Swedish naturalist of the mid-eighteenth century, who founded the modern system of biological classification. Under his rules, every species is known by a two-word Latinate name, announcing also its genus, and is classified within a hierarchy of nested categories. Below the level of kingdom (plant or animal), Linnaeus specified four other levels: class, order, genus, species. Later taxonomists, including Darwin, have parsed the living world more finely, into seven main levels—kingdom, phylum, class, order, family, genus, and species—plus a number of mezzanines (such as suborder, superfamily, subspecies) in between. Defining a hierarchy of categories, though, was only the most obvious part of devising a classification system. Two other questions were trickier: What causal reality (if any) did the categorical arrangement reflect; and how should a taxonomist determine which species to place where?

Linnaeus divided the animal kingdom into six classes, one of which was Vermes, encompassing not just earthworms and tapeworms and leeches and flukes but also sea cucumbers, slugs, snails, starfish, sea urchins, corals, bryozoans, octopuses, squids, oysters, and all the other mollusks, echinoderms, and crustaceans, including barnacles. It was a big bucket, his Vermes class, overflowing with homely, diverse critters. The next major advance came in 1795, when Georges Cuvier dissented from Linnaeus in a work whose title translates to *Memoir on the Classification of the Animals Named Worms*. Instead of lumping all those wormy and not-so-wormy invertebrates together, Cuvier split them into six new classes. In a later book he organized the various classes of animal into his four great *embranchements*, or phyla: vertebrates, mollusks, articulates (including those later known as arthropods), and radiates (circular animals such as starfish and sea urchins). He also made the case that each *embranchement* reflected a fundamental body plan, utterly distinct from the other three. The core of each pattern was the nervous system. All other anatomical attributes were functional modifications, according to Cuvier, suitable for particular conditions (life within a certain environment) and built around one of the four archetypal nervous systems. The existence of four *embranchements* was something Cuvier took as a given. Furthermore, he believed, the functional interdependence of organs was so intricate that one organ couldn't change without throwing the others out of whack. In other words, his system incorporated the idea of adaptation (arising from circumstances) and excluded the possibility of transmutation.

Cuvier's colleague, sometime friend, and implacable rival on matters of comparative anatomy was Etienne Geoffroy Saint-Hilaire, also working in Paris during the early nineteenth century. In contrast to Cuvier's functionalism, Geoffroy argued a formalist view. That is, he considered the form of a species to be deep-seated and basic. Diversity among species arose as peripheral and contingent modifications of archetypal forms, not as necessary functional attunements to external conditions. It might sound like only a difference of emphasis, but it was more. Beneath the multiplicity of animal shapes, Geoffroy discerned "unity of plan." The vertebrate skeleton, for example, was the template of one plan, providing a structural framework common to mammals, birds, reptiles, and fishes. This went well beyond Cuvier's notion about one sort of nervous system underlying each *embranchement*, with functional demands accounting for the varied anatomical details. Structure played the major role in dictating function, according to Geoffroy's view, rather than vice versa. The underlying structural plan was the main determinant of animal anatomy, to which adaptive modifications were secondary. Geoffroy admitted a possibility of some transmutation within lineages, but he didn't accept the idea of common descent for all creatures. Eventually he proposed an even broader unity, claiming that articulates belong with the vertebral group. An insect, he figured, is just a vertebrate wearing its skeleton on the outside. In 1830 he tried stretching the group still further, to include mollusks as well as articulates, on the evidence of supposed parallels between the anatomy of a cephalopod and of a vertebrate bent back on itself. But Cuvier, in a celebrated debate, shot that octopus out of the sky.

Another classification scheme that suggests the range of perspectives in those years was the quinarian system, a numerological approach dreamed up by a British entomologist and diplomat named William Sharp MacLeay. MacLeay's quinarianism was passingly influential. Robert Chambers for one had absorbed it, and he featured it in *Vestiges of the Natural History of Creation* when the book first appeared in 1844. Richard Owen, as a young lecturer, had drawn heavily from

MacLeay. Darwin too had read MacLeay (whom he knew through the Zoological Society) and, after a period of strong interest, reflected in the early notebooks, had rejected his notions. As MacLeay saw the living world, species were ordered by their similarities into circular sets of five. A circle would include five finch species, five iguana species, five whatever. Each species stood adjacent to similar species in a progression of resemblances around the circle, which closed on itself, the fifth species resembling the first. Every other taxonomic level also comprised five groups, again circularly arranged according to their degrees of resemblance. The animal kingdom consisted of five classes, for instance, corresponding to Cuvier's four *embranchements* plus another, Acrita (the amorphous ones, such as sponges). The class of vertebrates contained five orders. MacLeay's head was abuzz with fives. He seemed to believe that God's head was, too.

And the circles, the fives—they were just the start of it. MacLeay also saw affinities, parallelisms, analogies, osculations, all interconnecting the whole system like a bowl of magnetic Cheerios. Parallel to the animal kingdom was another circle constituting the plant kingdom, MacLeay asserted, though admitting his total ignorance of botany. Didn't it stand to reason that, if animals were quinarian, plants would be too? The circles at each level were linked to one another by what he termed *aberrant* groups or species, intermediates that didn't fit neatly into one circle or another, and by *osculants*, wherein two circles kissed. The platypus was an aberrant species somewhere between mammals and birds. But did it belong to both circles, or to neither? Answer: It was osculant. *Affinities* were the similarities among groups within a given circle. *Analogies* were the correspondences between groups in parallel circles. The affinity between reptiles and birds showed in hawksbill turtles. Tunicates were osculant between the classes Mollusca and Acrita. Barnacles were intermediates in the limbo connecting radiates and articulates—that is, between sand dollars and lobsters. It might all sound flaky to us but, in its time, quinarianism represented a careful attempt at finding order within biological diversity. In 1839 MacLeay went off to Australia, leaving his system to linger in the thinking of British biologists.

For Darwin, quinarianism was another source of headache, and he had plenty of those already. What frustrated him about all these systems was that they didn't consider *why* species might resemble one another, either in deep structural plan (for instance, all vertebrates) or in external particulars (for instance, two species of barnacle within the same genus). The problem with current taxonomic practice, as he had complained to George Waterhouse back in 1843, "lies in our ignorance of what we are searching after in our natural classifications." Linnaeus himself had admitted being clueless on that. "Most authors say it is an endeavor to discover the laws according to which the Creator has willed to produce organized beings," Darwin wrote. But to him those were "empty high-sounding sentences." He confided to Waterhouse his own wild view, which I quoted earlier, that classification should reflect "consanguinity," meaning descent from common ancestors. It wasn't metaphysics. It was genealogy.

The dull task of describing species, naming them, classifying them into an orderly system, was for Darwin an exercise in applying the idea of transmutation. What better enterprise for a man who wasn't yet ready to trumpet the idea itself? He'd write a fat monograph on Cirripedia, making it encyclopedic, straight-faced, inoffensive, even soporific, with transmutation as its implicit subtext. And if the result seemed more rational, more persuasive than other systems, then barnacle system-atics would constitute an elegantly laconic confirmation of his theory. That was worth some considerable measure of his energy and time, wasn't it? Yes . . . though not necessarily eight years.

15

His barnacle contact at the British Museum was John Edward Gray, keeper of the zoological collections, who in 1848 gave Darwin a scare. By then Darwin had committed himself to the barnacle project, with encouragement from Gray among others, and invested considerable work in it. Gray had helped persuade the museum trustees to allow Darwin an unusual privilege: borrowing all their barnacle specimens, receiving them batch-by-batch at his home, and snipping some into bits. Though he was intrigued by barnacles himself, Gray had even set aside his own research plans—so it seemed, anyway—in favor of allowing the field to Darwin. Then in March of that year, out of the barnacle blue, Gray presented a couple of short Cirripedia papers at the Zoological Society. Darwin let that much pass; either word didn't reach him (which is unlikely) about Gray's new barnacle efforts, or he didn't care, or he stifled his jealousy. Several months later, he heard a rumor from gossipy friends that Gray "intended to anticipate" Darwin's work, publishing descriptions of the strangest, most interesting barnacle species before Darwin could get around to them himself. This sounded like a dastardly deed—scientific poaching on his barnacle turf. He confronted Gray in person (which must have entailed a trip to the museum and a curt conversation, exactly the sort of event that often made him sick) and then followed the meeting with a huffy, legalistic letter. He would never have undertaken such a huge, tedious task, Darwin wrote, if he'd known that Gray was likely to swoop in and cherry-pick the juiciest barnacle revelations. Darwin apologized for having to raise the subject, but he didn't *sound* apologetic. Gray backed off, leaving barnacles to Darwin, and the museum's specimens continued to travel.

It was a small disagreeable moment that passed quietly, and almost wouldn't be worth mentioning, except that it illuminates a point of character. Did Darwin care about priority? Yes he cared about priority. The tiff with Gray suggests, alongside the earlier contretemps with Robert Grant over *Flustra foliacea* and a more distressing episode (with Alfred Russel Wallace) to come, that Charles Darwin possessed his reasonable share of human pride, and that he enjoyed not just the pure satisfactions of scientific inquiry and discovery but also the pleasures of publication, credit, renown, and getting there first.

Another sort of anguish hit him soon afterward, when his elderly father began to fade. Dr. Darwin was eighty-two, gouty and blimpish and confined to a wheelchair at the house in Shrewsbury, where Charles's unmarried sisters, Susan and Catherine, shared the nursing and fetching chores with two faithful servants. Charles had a complicated relationship with this intimidating man, the only parent he'd known since age eight. Dr. Darwin (or as Charles sometimes called him, "the Governor") was always generous in paying bills and dispensing pointed advice, but he had mostly delegated, to Charles's various sisters, the softer tasks of child rearing when Charles was a boy and the letter writing in years since. Charles respected his father's force of personality, his business acumen, and his powers of character judgment, and felt deep gratitude for his financial beneficence; but there were also a few scars on Charles's psyche, left by his father's sometimes bluntly expressed disapproval. He never forgot the remark about being "a disgrace to yourself and all your family" and would quote it (or reconstruct it from sulky memory) in an autobiography written many years later, when he was a codger himself. But in the same passage he would call his father "the kindest man I ever knew" and express loving forgiveness about that hard, unjust comment. Their relationship had improved as Charles became more successful, first as a lionized young naturalist who had capitalized well on his *Beagle* opportunity, then as a published author and respected scientist. The doctor and he seem to have worked through their difficulties, converting conflict to affection pretty well for a pair of bullheaded and stiff Victorian gentlemen. Also, Dr. Darwin probably thought better of Charles over the years as he saw that his younger son was the achiever in the family, whereas Erasmus had ripened into a full-time professional slacker, never marrying, nor practicing medicine, nor choosing to clutter his social schedule with any other sort of gainful employment. Charles didn't hold a job either, but he produced books, some of which earned good reviews and money. An additional bond between son and father was Charles's sense of guilt. Having withdrawn to his reclusive life in Kent, having centered his attention narrowly—on his work, on Emma and their children, on his own state of bad health—Charles knew he'd left his father and two unmarried sisters without as much of his presence and emotional support as they might have expected. The house in Downe represented voluntary solitude, while the house in Shrewsbury was plain lonely. And now, to the exacerbation of all these tangled feelings, Dr. Darwin was dying.

Charles last saw his father during a visit in October 1848. It was a period when, according to his diary, he himself had been "unusually unwell, with swimming of head" and also "depression, trembling—many bad attacks of sickness." He went to Shrewsbury anyway. His sisters presumably tended both patients. Back in May, during an earlier visit, there had been cheerful card games with the doctor, but probably not this time. Charles wrote another long letter about barnacles, to a scientist at Harvard, and otherwise occupied himself somehow in the gloomy manor. Then he came home. Three weeks later Catherine, the youngest of the family, wrote to say that Dr. Darwin was worse. We wheel him into the greenhouse, she explained, and he sits gasping for air, unable to speak above a whisper, but placid and ready. He tried to talk about you this morning but was overcome with emotion. Sorry your stomach is still bad, she wrote, reminding Charles of his self-absorption. The next day's letter from Catherine said: Father died this morning.

The funeral will be Saturday, she added, which should give you time to be here.

It did and it didn't. For one reason or another (he may have been too sick, or waiting for Emma to return from a visit), Darwin let a few days pass before setting out. Finally he went up to London on Friday, traveling alone, leaving his wife behind with their latest newborn and the house full of older children, servants, and barnacles. He stayed that night at Erasmus's place in the city, though Erasmus himself had already gone to Shrewsbury. During the solitary stopover, grieving for himself as well as for his father, he scribbled a note back to Emma that shows the depth to which his loving dependence on her had grown:

My own ever dear Mammy.—
Here I am & have had some tea & toast for luncheon & am feeling very well. My drive did me good & I did not feel exhausted till I got near here & now I am rested again & feel pretty nearly at my average.—
My own dear wife, I cannot possibly say how beyond all value your sympathy & affection is to me.—I often fear I must wear you with my unwellness & my complaints.
Your poor old Husband, C.D.

The next day he caught a train onward to Shrewsbury, arriving at the family home after the cortège had rolled away toward the church. Rather than chasing it, Charles stayed at the house, along with one married sister, who like him didn't feel up to enduring the funeral or standing at graveside. Later he rationalized that it was "only a ceremony," but admitted he felt sad at the "deprivation." No one will ever know whether he had dawdled on purpose to miss seeing his father buried. But his absenteeism from grim ceremonies and big events was more than accidental. It was a pattern.

16

When his health allowed, Darwin was seeing some exciting things among the barnacles: bizarre kinds of intermediate sexuality, homologies with other crustaceans, rudimentary or "abortive" structures, barnacle noses, barnacle ears, a larval stage with no mouth, and cement glands that vent through the antennae to form a point of attachment when an immature barnacle finally settles down to adulthood, gluing itself forevermore onto a rock, a log, or the hull of a ship. Species of the genus *Proteolepas*, Darwin noticed, have no legs. Species of *Alcippe* develop only three body segments of the seventeen possessed by an archetypal barnacle, omitting fourteen, and as though that's not enough, the female has no anus. Species of *Ibla* include only females—or so it seemed, until he looked closer and found tiny parasitic males, hardly more than sperm capsules, embedded in one female's flesh like blackheads. All of these oddities were particular to a genus, or to one or more species within a genus. He also found notable differences on another categorical level: *within* species. Contrary to what he'd believed all along about the rarity of variation in the wild, barnacles turned out to be highly variable. A species wasn't a Platonic essence or a metaphysical type. A species was a population of differing individuals.

He wouldn't have seen that if he hadn't assigned himself the tricky job of drawing lines between one species and another. He wouldn't have seen it if he hadn't used his network of contacts and his good reputation as a naturalist to gather barnacle specimens, in quantity, from all over the world. The truth of variation only reveals itself in crowds. He wouldn't have seen it if he hadn't examined multiple individuals, not just single representatives, of as many species as possible. "I have been struck," he wrote Hooker, who had asked for a barnacle update, "with the variability of every part in some slight degree of every species." This individual had a bigger penis, or shorter legs; that individual had a longer stalk, or a wider thorax. The project involved more nuances and judgment calls than he had expected. How much variation is too much to lump within a single species? What distinguishes the species category itself from the category of varieties (such as breeds of dog) within a species? A person could go blind and crazy. "Systematic work wd be easy were it not for this confounded variation," he told Hooker, then admitted it was "pleasant to me as a speculatist though odious to me as a systematist." The speculatist in him was thinking about transmutation, not just looking for taxonomic order. Abundant variation among barnacles filled a crucial role in his theory. Here they were, the minor differences on which natural selection works.

But all this gathering and examining, dissecting and describing, consumed month after month when he was well enough to work, and oppressed him even more when he wasn't. In March 1850, almost four years into the effort, he grumped to Lyell: "My cirripedial task is an eternal one; I make no perceptible progress." That was weary exaggeration but it reveals how he felt. So does the comment several months later, to Hooker, that at last he was sending to the printer "a small, poor first fruit of my confounded cirripedia." By then he'd decided that his monograph would be a four-volume work, two volumes on the stalked barnacles (one for the fossil forms in Britain, one for the living forms worldwide) and a corresponding pair (fossil forms, living) on all the others. The "small, poor first fruit" was his little volume on the stalked fossil forms, an arcane work for a limited audience, published by the Palaeontographical Society in 1851. The volume on living species followed later that year, also released quietly by a specialty publisher. Immediately he went on to the sessile barnacles. A year later he reported to his old friend W. D. Fox that he was still at work on the Cirripedia, "of which creatures I am wonderfully tired: I hate a Barnacle as no man ever did before, not even a Sailor in a slow-sailing ship." The barnacle project was now mirroring his *Beagle* experience: a long lonely voyage, which might or might not pay off.

It did pay, to some degree anyway—not just in private scientific insights but also in public acclaim. Two years after the publication of his stalked-barnacle volumes, the Royal Society awarded him their Royal Medal for Natural Science in recognition of that work. Hooker sent him the news, describing the decisive Society meeting at which Darwin's name was put in nomination, followed by "a shout of paeans for the Barnacles" that would have made the barnacle man smile. Several weeks later, Darwin actually went up to London to receive the award. Maybe he did smile. If so, it was just for a few minutes in public, before the excitement brought back his vomiting.

The Royal Medal was "quite a nugget," he mused privately. Bestowed for such a dry, conventional enterprise as barnacle taxonomy, it gave him a big new measure of scientific esteem and stodgy credibility. He had been garlanded by the establishment. With what was coming, he knew, he would need every leaf.

17

But the smile and the gold medal were unimaginable back in early 1849, as he tried to resume work on the stalked barnacles after his father's death. His symptoms had returned. "Health very bad with much sickness & failure of power," he told his diary. He limped along until March, then took desperate action. He packed up Emma, the children, the butler, the governess, and several maids, and went off to a water-cure establishment run by Dr. James Gully in the town of Malvern, in Worcestershire, near the Welsh border. It was a two-day journey by train and coach, a major logistical operation for such a group, including five young kids and a screeching infant (the latest little son, Francis). You had to be genuinely sick to think that this might help. Darwin had heard about Gully from friends and read the doctor's book full of quackish ideas, *The Water Cure in Chronic Disease*. For lack of other options, he was ready to take the plunge.

The theory behind Gully's so-called cure was that excess blood, congested in vessels servicing the stomach, caused "nervous dyspepsia" such as Darwin's. The solution to that, Gully thought, lay in drawing blood away from the stomach to the skin and extremities by means of cold water and friction, generating just enough chilly irritation to produce a rash. Wrapping the body in wet sheets had the additional benefit of lowering brain function, which also helped ease the stomach. Darwin was put on a daily regimen that included scrubbings with a cold wet towel, soaking his feet in cold water, drinking tumblers of cold water, wearing a wet compress against his stomach all day, heating himself into a sweat with an alcohol lamp before being rubbed again with cold towels, taking walks and naps between these torments, swallowing homeopathic medicines, and subsisting on a bland diet that excluded, as Darwin said, "sugar, butter, spices tea bacon or anything good." At first Gully allowed him a little snuff, six pinches per day, then made him quit that habit altogether.

Darwin grumbled about the diet and the snuff deprivation, and was skeptical of Gully's belief in homeopathy (not to mention Mesmerism and clairvoyance, two of the doctor's other enthusiasms). But he persuaded himself that the water-torture was working. Within the first eight days he was pleased to see an eruption of some sort broken out all over his legs. He went a month without vomiting, a notable stretch, and gained some weight. One day he even walked seven miles. "I am turned into a mere walking & eating machine," he told Fox. A side effect of the treatment, he reported cheerily to another friend, was "that it induces in most people, and eminently in my case, the most complete stagnation of mind: I have ceased to think even of Barnacles!" Gully led him along with guarded assurances that Darwin could be cured, but that it would take time. How much time? Always a little more. After three and a half months at Malvern, the family returned to Downe, but Darwin brought some of Gully's regimen with him. He ordered a showering hut built in the garden, with a raised cistern that could be filled from the well, and took a cold shower there around noon every day. Hours before that, first thing in the morning, he did a sweat session with the alcohol lamp, jumped into a cold bath, and then endured cold-towel scrubbing by Parslow, the faithful butler, who saw things during his thirty-five years with the Darwins that were almost as odd as barnacle sex.

Darwin felt much better. He went back to his dissecting scope. For exercise, he bought a horse and started riding. He made plans to go to Birmingham for the annual meeting of the British Association, nearly inescapable since he was now a vice president. The problem with Gully's cure, though, was its impermanence. At the Birmingham gathering, away from his alcohol lamp and cold showers, surrounded by social excitement and loud, pompous colleagues, he felt queasy again. Instead of going on a scheduled field trip, he skittered down to Malvern for a tune-up with Gully. Back home, he continued the water treatments and, to avoid overexertion, gave up all reading except newspapers. He allowed himself just two and a half hours each day on barnacles, spending much of his time cold and wet. No wonder the work went slowly.

Within the next couple of years he returned to Malvern twice more. The first trip was another refresher visit for himself, a nice week in June, during which he divided his attention between wet towels and barnacles, and claimed that he'd gotten to like his own "aquatic life," except for the incessant dressing and undressing. This time he didn't congratulate himself for mental stagnation. His brain was alert, the monograph seemed to be progressing, people told him he looked well, and in an upbeat letter to Hooker he made that comment about the mixed thrill of finding so much "confounded variation" within species of Cirripedia. The second trip was very different. He brought his oldest daughter, who was now mysteriously ill herself.

Annie Darwin was ten, a bright and generous-hearted little girl with a special bond to her father. He loved her joyousness, he admired her goodness, he treasured her company. He confided to Fox that she was his favorite child. He would indulge her sometimes to spend half an hour arranging his hair—making it beautiful, she said—or fussing with his collar or cuffs. She would sneak him pinches of snuff when he was supposedly abstaining. She would dance along when he strolled the Sandwalk. "Her whole mind was pure & transparent," Darwin wrote later.

At age eight, Annie had suffered through scarlet fever, a life-threatening disease, but seemed to recover. Or maybe not quite. Six months later, her mother started noticing that she wasn't right. Something came over Annie's buoyancy like a cold afternoon shadow. She was fretful and intermittently feverish; she cried often, especially at night. They sent her to the resort town of Ramsgate, with the governess and her little sister Henrietta (known as Etty), for sea air and shell-collecting on the beach. They sent her to an eminent doctor in London. They bought her a canary. Nothing helped. Around Christmas, she started to cough. Darwin worried that she had inherited his "wretched digestion," not knowing it was worse than that. On the misguided guess that nervous dyspepsia was troubling her girlish tummy, early in 1851 they put her on Dr. Gully's water-cure regimen. With cold water from the household well, Annie got the same treatment as her father: wrapping, rubbing, foot-soaking, and frigid baths. She caught influenza, seemed to recover from her flu, but still wasn't well. The cough lingered. Some days were better than others. In late March, Darwin packed her off to Malvern for the full Gully experience.

Unidentifiable fevers were common in those years before the discovery of pathogenic microbes, when educated people still thought malaria was caused by miasmal vapors from swampy land and no one knew a virus from a hangover. Annie Darwin's illness resembled her father's in one sense: It was never conclusively diagnosed. A modern scholar named Randal Keynes (with a good feel for the family history and special access to some sources, being himself a great-grandson of Annie's brother George) has made a compelling try at solving the riddle in retrospect. Keynes presented all the available evidence to four medical historians and asked for informed speculation. Their consensus was that Annie probably suffered from tuberculosis, which sometimes attacks the brain, the intestines, or other organs as well as the lungs. It was a dreaded killer in the nineteenth century, familiar as "consumption" or "phthisis" but not well understood, a bacterial disease that seemed to move like the angel of death. There was no cure (until antibiotics were developed), and if there had been a cure, dousing the patient in cold showers and wrapping her with wet towels wouldn't have been it.

But they didn't know. Darwin escorted his cherished daughter to Malvern and left her there, with the children's nurse, the governess, and little Etty for company, to Gully's drenching care. After two weeks, Annie started to vomit, then fell into another fever and grew weak. Gully thought that she had passed through a modest crisis and would improve. She didn't. Darwin was back home at Down House, probably hunched over his barnacles, when word arrived from Malvern that he'd better come quick. Emma was pregnant again, eight months along, so he went by himself, immediately.

Annie's condition over the following week is vividly documented in Charles's letters to Emma, which reached Down House by overnight delivery. On Thursday, the little girl looked poor but "her face lighted up" at sight of her father. On Friday, her pulse steadied but she vomited badly, and "from hour to hour" she seemed to be tangled in "a struggle between life & death." Emma's own return note that morning, whatever it said, made Charles cry. Next day, Annie's "hard, sharp pinched features" left her almost unrecognizable, but the fever was gone, and she sipped some gruel. Sunday was Easter, a fact of small interest or importance to Darwin, who didn't mention it in his account of Annie's continual retching. She hadn't lost her sweet spirit, he reported; given a gulp of water, she said weakly, "I quite thank you." And so on. It was a week of grim pathos. She died at noon on Wednesday.

The scene around Annie's deathbed, in the hours following, was chaotic as well as doleful. The governess immediately had "one of her attacks," according to Charles and Emma's sister-in-law, Fanny Wedgwood, who had come to Malvern to lend some support. The nurse was desolate and useless, too. As for Charles, "I am in bed not very well with my stomach," he told Emma, at the end of his short note announcing Annie's death. "When I shall return I cannot yet say." He was confused and exhausted and depressed, as well as sick again, and in some small degree also relieved that Annie's suffering had ended. "She went to her final sleep most tranquilly, most sweetly," he wrote, trying to offer Emma some consolation. He mentioned God three times within the space of a page, allowing himself the rhetoric of conventional piety. "I pray God," so he claimed, that an earlier note from Fanny had prepared Emma for the bad news. It's unlikely that, in a literal sense, he had prayed God on any such point. "God only knows" what uglier miseries Annie might have suffered during a longer life. "God bless her," he said simply. These comments are curious in light of his own disbelief (which solidified around this time) in a benevolent Christian deity. Annie's funeral was set for Friday.

He skipped it. Early on Thursday morning he grabbed some books, left behind his extra clothes, and caught a train for London. Making good connections, he was in Downe by evening. His excuse for this hasty getaway was that Emma needed him now more than Annie did, and that it would be soothing to his wife, in her delicate condition, if they could weep together. Maybe that was indeed the driving reason, and maybe not. Emma, for her part, had written back that he shouldn't feel a need to hurry. But she'd agreed on one point: "We shall be much less miserable together." Despite the theological disagreement, they were now deeply connected as lovers, partners, parents, and each other's main source of emotional support in hard times. The only other human quite so dear to him had been Anne Elizabeth Darwin.

On Friday, April 25, 1851, Fanny Wedgwood and her husband (Emma's brother), along with the governess and the nurse, rode to the Malvern churchyard behind a hearse carrying Annie's coffin. The nurse was so dispirited she had to be lifted into the carriage; the governess was more composed now, crying only intermittently. It was a small party of mourners. The younger sister, Etty, had long since been shipped away to other relatives. Dr. Gully didn't attend. Darwin himself sat at home that day, writing Fanny a letter of thanks for encouraging him to leave Malvern and for handling the funeral arrangements. "Sometime," he added, he would like to know in which area of the church graveyard Annie's body was buried.

The bare facts make him sound callous. He wasn't. His emotions were dark and deep. But apart from his devotion to Emma, and his dependence on her, he had a strong instinct of self-protection. And he was now closing himself up like a barnacle.

18

The death of Annie in 1851, following the death of his father three years earlier, marks an important point in Darwin's long, quiet disengagement from religious belief and spirituality. He avoided both funerals and left the requiescats to be spoken by others, not just because some physical or emotional weakness made him think he was unable to stand in black beside a coffin; he also seems to have considered those Anglican burial rites, with their assurances of resurrection into eternal life, false and pointless. Years later he told a pair of radical philosophers, who had begged an audience with him while they were in London for a freethinkers' convention, and who were accommodatingly invited down for lunch, "I never gave up Christianity until I was forty years of age." His fortieth birthday had fallen between the two deaths.

What caused his loss of conventional religious faith, and how far he eventually shifted toward atheistic materialism, are complicated questions. To the two freethinkers, he said drily that Christianity was "not supported by evidence." In his autobiography, he wrote that "disbelief crept over me at a very slow rate, but was at last complete." It crept so slowly, in fact, that he "felt no distress," and had "never since doubted even for a single second that my conclusion was correct." One contributing factor to his apostasy was the careful reading he'd done on philosophical and scriptural topics, ranging from Hume, Locke, and Adam Smith to James Martineau's *Rationale of Religious Enquiry*, and from Paley, Herschel, and Ray to John Abercrombie's *Inquiries Concerning the Intellectual Powers and the Investigation of Truth*. He took some interest in the works of Francis Newman, a Latin professor whose elder brother, John Henry, had turned Catholic and would eventually become Cardinal Newman; Francis Newman's spiritual journey went in the opposite direction, toward austere and skeptical Unitarianism. Darwin read Newman's *History of the Hebrew Monarchy*, which critiqued the Old Testament for its dubious historicity, as well as his autobiography, *Phases of Faith*, and still another Newman book, *The Soul, Her Sorrows and Her Aspirations*, billed provocatively in its subtitle as a "natural history." Such influences complemented Darwin's own empirical disposition. He rejected the Gospels as revealed truth, the notion of eternal punishment for unbelievers (such as his own father and grandfather), the immortality of the human soul, Christian theology in general, and Paley's old argument for the existence of an immanent personal deity from the evidence of clockwork nature. Special creation? Divine providence? Godly design? Darwin had found no support for those notions in biogeography, the taxonomy of barnacles, or the fates of certain innocent children. "Everything in nature," he concluded coldly, "is the result of fixed laws." Had an impersonal First Cause of some sort, a Supreme Being in the fuzziest sense, given rise to the universe and set it in motion according to the mechanics of those fixed laws? Maybe. For much of his adult life, including the period when he wrote *The Origin of Species,* that's what Darwin felt inclined to believe. Later, "with many fluctuations," he grew gradually more doubtful. It was impossible to know. The best way of describing his spiritual convictions or lack of them, Darwin declared in the autobiography, was to label him an agnostic.

Elsewhere he simply called himself "muddled" on these big, irresolvable issues. He was bothered particularly by two contradictions he saw in or around orthodox Christian dogma: the tension between a law-governed universe and an intervening God, and the occurrence of evil in a world designed by an omnipotent deity who prefers good.

Did physical laws encroach on divine prerogatives? For some thinkers they did, and not just in the biological realm. Even Newton's law of gravitation, Darwin knew, had once been attacked by Leibniz as "subversive" of natural religion. Gravity was "an occult quality," a godless fudge factor, invoked wrongly to explain the miraculous orbiting of planets—or so went Leibniz's complaint. Had that criticism been accepted by reasonable people? No. Mostly they preferred Newton's nice, grounding law. Then why accept the same complaint when applied to diversity and adaptation among living creatures? "I cannot believe that there is a bit more interference by the Creator in the construction of each species," Darwin wrote, "than in the course of the planets."

The problem of evil, and of gratuitous suffering inflicted on innocents, troubled him at least as much. "I cannot see, as plainly as others do," he wrote to Asa Gray, an American botanist friend at Harvard, "evidence of design & beneficence on all sides of us. There seems to me too much misery in the world." Why would a benevolent God design ichneumon wasps, for instance, with the habit of laying eggs inside living caterpillars, so that the wasp larvae hatch and devour their hosts from inside out? Why would such a God design cats that torture mice for amusement? Why would a child be born with brain damage, facing a life of idiocy? he asked. Several months later, again to Gray, he pressed further: "An innocent & good man stands under [a] tree & is killed by [a] flash of lightning. Do you believe (& I really shd like to hear) that God *designedly* killed this man? Many or most persons do believe this; I can't & don't." He wasn't arguing only about a hypothetical man and a hypothetical flash of lightning. He was drawing on personal experience: the problem of evil as revealed in watching a ten-year-old daughter die of some wasting illness. Any god who controlled events on Earth closely enough to preordain such an occurrence—or to permit it, if divine permission was necessary—wasn't one that Darwin could take seriously.

A week after Annie's death, while the images were fresh, he wrote a short private memoir recording a few of her charms, habits, and traits. Her pirouettes along the Sandwalk. Her prissy neatness. Her affection toward the younger children. Her talent for music. Her enthusiasm for dictionaries and maps. He and Emma had lost the joy of their household, Darwin wrote, and the solace of their old age. Surely the little girl must have known how much she was loved. "Blessings on her," he ended, vaguely, this time omitting the name of God.

A Duck for Mr. Darwin

1848–1857

19

Darwin was busy at his barnacles, back in April of 1848, when a young man named Alfred Russel Wallace left Liverpool aboard a ship bound for Brazil. The two weren't personally acquainted, not at that time, and Wallace (like the rest of the world) was ignorant of Darwin's secretive work on transmutation. But he wasn't oblivious to the subject. Wallace knew just enough natural history to be dissatisfied with the old explanations of species diversity, its distribution and origins. He wanted something more than natural theology. Now he was headed to the tropics in search of adventure, rare birds, gorgeous butterflies, giant beetles, and a chance to contribute new facts—maybe even insights— toward what he called "the theory of the progressive development of animals and plants."

Wallace's main source of information about that theory, the book that had inflamed his enthusiasm, was *Vestiges of the Natural History of Creation*, by then in its seventh edition. Unlike those critical readers who dismissed *Vestiges* as junk,

Wallace had found it a provocative starting point. At the book's core he saw an ingenious hypothesis. This calls for further research, he concluded. This calls for me to jump on a boat to the Amazon. He was twenty-five years old, bright and ambitious, impetuous and impressionable, scientifically untrained. Events would show that he was also persistent, observant, and tough.

Besides being fourteen years younger, Alfred Wallace differed from Charles Darwin in a whole list of ways: no family wealth, no university education, no mentoring by Anglican naturalists, no social connections to the British Navy, no opportunity to travel the world as a relatively pampered guest aboard one of Her Majesty's ships. Wallace was the eighth child of nine born to middle-class parents who lacked middle-class income. His father, trained as a lawyer, had a knack for bad investments and a disinclination to practice law, so the family was downwardly mobile. Alfred left school at age fourteen, when the money for his childhood ran out, and apprenticed as a surveyor. For much of the next decade he surveyed railroad routes and property boundaries across the landscape of England and Wales, living out of inns, boarding houses, sometimes a rented cottage, while catching what education he could at mechanics' institutes (self-improvement facilities for laborers) and public libraries. He had always loved to read. For a young surveyor, too curious and driven to piss away his evenings in pubs, the institutes and libraries opened vistas. He read Alexander von Humboldt's great narrative of travels in South America (which had also inspired Darwin), William Prescott's *History of the Conquest of Peru*, Lyell's *Principles of Geology*, William Swainson's *A Treatise on the Geography and Classification of Animals* (which described MacLeay's quinarian systematics), and John Lindley's *Elements of Botany*. He read Darwin's *Journal* twice and found it thrilling, second only to Humboldt's as a science-flavored travel narrative. He read W. H. Edwards's rollicking new book, *A Voyage up the River Amazon*. He read Malthus.

Meanwhile he developed a passion for life outdoors, hiking transects across the Welsh mountains, and began making himself into a naturalist. His early efforts focused on botany, until a new friend turned his head toward beetles. The friend was Henry Walter Bates, an apprentice in the hosiery business but restless for getaway and keen on natural history, like Wallace. They met in Leicester during a year Wallace spent there, on a temporary escape from surveying, as a schoolteacher. When he saw Bates's collection of beetles—shiny like jewels, amazingly diverse, and almost all found near Leicester—Wallace was hooked. He got himself a collecting bottle, some pins, and a beetle box, and laid out precious shillings for a *Manual of British Coleoptera*. Bates helped him learn where to find beetles and how to identify them. When the teaching job ended and Wallace went back to Wales, he and Bates stayed in touch, sharing thoughts about scientific books and trading specimens of rare British beetles. Have you read *Vestiges*? he asked Bates in one letter. Have you read Lawrence's *Lectures* on comparative anatomy? William Lawrence had been one of the radical materialists teaching anatomy in London, a subversive influence even before Robert Grant arrived. Wallace found his book "very philosophical," meaning solid in logic and scientific method. Its discussion of variant races among the human species, he told Bates, bore directly on what interested Wallace so much: the theory of progressive development. He was already doubtful, as revealed in this letter to

Bates, that the distinction between species and varieties was so clear and absolute as most people assumed.

Around that time, probably while Bates was in Wales for a beetling visit, they conceived the idea of a more daring excursion. They'd go to the Amazon together, paying their expenses by shipping back natural history specimens for sale to dilettantish collectors in England. This wasn't as impractical as it sounds; in that era, some gentleman dabblers kept cabinets of small biological trophies for display, just as others might show off their French paintings, their Chinese ceramics, or their items of native art. Wallace and Bates lined up a London sales agent, one Samuel Stevens, who knew the retail trade in beetles and butterflies. They equipped themselves with guns, nets, and other field gear, and finagled a few letters of introduction. Wallace got vaccinated. Bates was with him when the boat anchored at Pará, the Brazilian port near the mouth of the Amazon, on May 28, 1848.

Neither of them could have said just where this scientific lark was headed or how long it would last. At one point Wallace told Samuel Stevens that he hoped to be back in England by Christmas of 1850. Instead he roamed the Amazon basin for four years. Bates stayed for eleven.

After some months of collecting side by side, and mostly near the Amazon's mouth, they separated in order to follow different instincts and to minimize competition. Wallace headed upriver. He worked to learn Portuguese and the Indian trade language. He shot birds, skinned them, and defended the skins against jungle rot and voracious ants. He collected gaudy butterflies and glittery beetles. He caught fish and pickled them in spirits. He carefully packed these various specimens into crates for shipment to England, adding other items (a small stuffed caiman, a pair of Indian calabashes) to fill out a crate. From his collected specimens, and from the landscape scenes around him, he made notes and sketches. He drew maps. He was curious about everything, human cultures and tropical vegetation as well as collectable animals. He recorded some anthropological observations and made a small study of the diversity and practical uses of palm trees. Eventually he ascended to the Rio Negro, a vast blackwater tributary of the main Amazon, and explored its upper reaches by canoe for most of two years.

From a branch of the upper Negro he set off on an overland hike to the Serra do Cobati, a stony massif rising out of the forest, in search of a creature known as *o galo-da-serra*, the cock-of-the-rock. This bird, blazing vermilion except for its wing feathers and tail, with a disk-shaped crest obscuring the front of its face, was a spectacular oddity well worth the ten-mile slog. Two species are recognized nowadays. The one Wallace saw, the Guianan cock-of-the-rock (*Rupicola rupicola*), lives only around eroded mountain outcrops in the jungles of eastern Colombia, Venezuela, and northern Brazil, where the females build mud nests in crevices amid the steep rock. Both sexes feed on fruit. The males compete for females by assembling in display areas, known as *leks*, and taking turns showing off their physical splendor. Wallace spotted one in a dim thicket, "shining out like a mass of brilliant flame." He raised his gun; the bird spooked and flew off; but after a little more tracking he got a second chance, and killed it. With help from his crew of Indian hunters, he eventually took twelve cocks-of-the-rock. Their lekking behavior, bringing them together in vulnerable crowds, may explain how he bagged so many.

Those dozen birds, packed into a small box for shipment

home, embody a critical aspect of Wallace's collecting career, both in the Amazon and later: redundant sampling. That is, he wanted quantity, not just diversity. Because he was paying his way on a piecework basis, and because *Rupicola rupicola* was extraordinarily decorative, he killed as many individuals of the species as he could. Darwin, a rich man's son collecting only for himself, might have taken just one or two. Wallace had hoped for fifty cocks-of-the-rock, but was glad to settle for twelve.

When he laid them all side by side, did he notice intraspecific variation? Did he find that not every individual was quite so luminously red as the others? Did he see that some were rather more orangish? Did he detect differences in the diameter of the facial crest, or in the width of the thin yellow band across the tail? Did he recognize from such differences that multiple sampling of a single species *isn't* actually redundant, but in fact yields information about variability? We don't know. He didn't say. But we can wonder. The abundance of naturally occurring variation within species was a crucial clue to the transmutation mystery, unnoticed by most naturalists of the day. Darwin needed eight years with barnacles, following five years of travel and ten years of study, to awaken him about variation in the wild. Wallace saw it sooner because, besides being an alert observer, he was a commercial collector, hungry and broke.

Not that his insight about variation, or all his data, came easily. Wallace paid high human costs for everything he gleaned from the Amazon. Beyond his sheer labor, there were hardships—loneliness, the danger of drowning or being murdered or snake-bitten, pestilential mosquitoes and sandflies, delays and frustrations in hiring helpers or finding supplies, cash shortages when his letters of credit from Stevens didn't arrive, days of subsisting on manioc flour and coffee, the incessant struggle to keep order among his collections and his thoughts, all within the relentless, gnawing entropy of a tropical forest. He spent two weeks with his arm in a sling, unable to work, because of an infected hand wound. His younger brother, Herbert, came out to learn the collecting trade at Alfred's side but retreated back to Pará when it didn't suit him, and then died there of yellow fever. Wallace himself was laid up by unidentified fevers more than once. He penetrated far into the headwaters of the Rio Uaupés, in what is now eastern Colombia, hoping to nab a rumored white form of a spectacular black bird, the umbrella chatterer, and was forced to conclude that the white version probably didn't exist. Early in 1852, satisfied or exhausted, he started down from the Rio Negro headwaters in an overloaded canoe.

He was bringing six crates of his specimens that hadn't yet been shipped, plus all of his journals and notes and drawings, plus a menagerie that he hoped he might wrangle back to England alive: five monkeys, two macaws, twenty parrots and parakeets, and some other birds. He reached Pará around the end of June, completing a loop he'd begun four years earlier, and visited Herbert's grave. On July 12 he boarded a ship, the brig *Helen*, bound for England.

The *Helen* was an ill-fated tub. After three weeks at sea, still in mid-Atlantic, she suddenly caught fire. A combustible hazard in the hold, casks of balsam, had spontaneously bubbled into flame, taking the captain by surprise. Wallace stumbled down into his smoky cabin and grabbed what he could, tossing papers into a tin box. From all his Amazon treasures he saved just a small bundle of drawings and some notes. He was forced to abandon the specimen crates, which included his private collection of insects and birds, as well as most of his written records. He climbed into a leaky lifeboat with other castaways and watched the *Helen* burn, then sink, dragging his charred diaries and roasted parrots to the bottom. Four years of work dropped away like a cinder in a bucket.

After ten days in the open lifeboat, patching its leaks with bits of cork, living on biscuit and raw pork and carrots, Wallace and his companions were rescued by another English ship, which turned out to be almost as blighted as the *Helen*. This old, slow boat, the *Jordeson*, nearly sank twice in high seas before they reached home. With two crews aboard, there wasn't enough to eat. The hold was overloaded with Cuban hardwood. Not far from England they were hit by a gale that split one sail and nearly defeated the bilge pumps. Almost three months after leaving Brazil, Wallace limped ashore at Deal, in southeastern England. His ankles were swollen, his legs were weak. He celebrated, along with the *Helen* and *Jordeson* captains, over a dinner of beefsteak and plum tart. They were glad to be alive and then, presumably, glad to go separate ways.

Wallace went to London. A less intrepid man, or a less stubborn one, would have written it all off as the misadventure of a lifetime and wanted no more. Not Wallace. Four days later he admitted to a friend that, though he had sworn never to sail another ocean, "good resolutions soon fade." He was already plotting his next expedition. He hadn't found a solution to the big mystery, the one about progressive development of animals and plants. He was more than ever convinced that such progression occurs, and that it could be explained by some physical process or law. He wanted a new arena for collecting and observing. Maybe he'd go to the Andes. Or to the Philippines. He had seen a great river. Now he might look at mountains or islands.

20

Despite all his bad luck, his near-death experiences and dispiriting losses, Wallace's four-year excursion in the Amazon produced some important rewards. It was his second apprenticeship; instead of teaching him the surveyor's craft, this one let him develop the skills and strengths of a tropical explorer, an expert collector and preserver of specimens, a sharp observer of animal diversity and other biological patterns. It began the process of alerting him to the significance of varieties within species. It further excited his thinking about progressive development. And it made him a biogeographer.

Biogeography, as I've mentioned, is the study of animal and plant distribution around the planet. It addresses two simple questions: Which kinds of creatures live where, and why do they live there but not elsewhere? Its significance for any theory of biological origins—an evolutionary theory, say, or a creationist theory—is that biogeography represents a complicated body of empirical facts that the theory must explain. Why do the Galápagos Islands harbor three endemic species of mockingbird, all closely related but no two of them native to any one island? Why do polar bears live in the Arctic, penguins in the Antarctic, and not vice versa? Why do tree kangaroos (arboreal marsupials of the genus *Dendrolagus*) inhabit tropical forests in northeastern Australia, and also in nearby New Guinea, but not tropical forests in South America or Africa? Why do hummingbirds and toucans occur only on one side of the Atlantic Ocean (in the Americas), while sunbirds and hornbills occur only on the other side (in Africa and farther eastward) of that ocean? A possible answer is that God specially created each species, plunking it into one ecosystem or another as suited His own opaque whim. This explanation is not wholly satisfying to the intellect, though it seems adequate to some people of faith. Another answer is that all creatures have evolved from common ancestors; that they have diverged slowly into distinct lineages and species, dispersing to new habitat as opportunity has allowed, although their dispersal has often been restricted by physical barriers such as mountains or seas; and that the current geographical distribution of species reflects the history of such divergence, restriction, and dispersal. That's the answer favored by Darwin after he had seen the Galápagos and the plains of South America. Wallace came to it by way of the Amazon.

From the ashes of Wallace's misfortune on the *Helen* arose one squawking phoenix of good news: His agent, the reliable Samuel Stevens, had insured those collections for £200. Maybe Stevens was prescient, or had once been a sailor himself, or had recently seen *The Merchant of Venice*. Anyway, that money accounts for how Alfred Wallace could now live as a man-about-London for more than a year, taking a role in the city's scientific societies, writing papers and books, rather than being forced back to surveying rail lines across the Welsh countryside.

Wallace attended meetings of the Entomological Society, including one just after his landfall, when he could barely walk. He went not as a member (there was still class bias against commercial collectors), but as a visitor sponsored by Stevens, who was an officer of the bug-lovers' club. Stevens had already made him famous to this group by excerpting his letters for publication and showing his shipped specimens, as John Henslow had done for Darwin fifteen years earlier. Darwin himself was a member of the Entomological Society but, hidden away at Downe, rarely attended its meetings. Those who did attend had seen Wallace's black-and-yellow swallowtail butterfly, *Papilio columbus*, and they may have read his report on the umbrella bird, as published in the *Annals and Magazine of Natural History*. Stevens also got him entree to the Zoological Society, where on December 14, 1852, he delivered a paper, "On the Monkeys of the Amazon."

This paper contained Wallace's first real statement on biogeography. Having sighted twenty-one different species of monkey along the Amazon and Rio Negro, he had noticed something remarkable. The species on one side of each big stem of river differed from the species on the other side. What he termed "closely allied species," such as two species of marmoset belonging to the same genus, were in some cases localized on opposite banks. The rivers themselves—the mainstem Amazon and its largest branches, the Negro and the Madeira, which form a vast chicken-foot pattern across the entire basin—seemed to constitute nearly impassable boundaries of distribution. North of the Amazon and east of the Negro lay one biogeographical district, which Wallace labeled the Guiana. Westward of the Negro lay the Ecuador district. South of the Amazon, the Rio Madeira delineated two more districts, called by Wallace the Peru and the Brazil. So far as monkeys were concerned, those four major districts of Amazonia, divided by wide stretches of moving water, might as well have been islands.

21

Islands: Maybe they could tell him something more. Biogeography: Maybe he should pay keener attention. Closely allied species: What did the patterns of their geographical distribution suggest? Wallace planned his next field journey with these considerations in mind. He worked his contacts at the various scientific societies, including the Royal Geographical, to get more letters of introduction and free passage on an outbound boat. After just a year and a half back in England, during which he published two unspectacular books (a small volume on palm trees and *A Narrative of Travels on the Amazon and Rio Negro*, which suffered lack of concrete detail because of the lost journals), he was set to go. This time, east. Early in 1854 he left on a Peninsular & Oriental steamship, making connections that put him in Singapore by late April.

Singapore, a bustling international port with just a small patch of forest nearby, only suited him briefly. It was sited on an island, which might have been good, but this island wasn't a remote and unexplored refuge harboring an abundance of wondrous, unknown species. It was a cosmopolitan crossroads. Also, Chinese woodcutters were whacking away at what forest remained, harvesting timber and planting vegetable gardens on the raw land. Using the woodcutters' trails, Wallace found plenty of nifty insects, especially among the beetles, but few birds or mammals. He tried basing himself at Malacca, a town farther north, on the Malay Peninsula, but after two months and another spell of fever he wanted to move again. He considered going up to Cambodia with a missionary he'd met, a congenial French Jesuit who spoke four languages. When the Jesuit delayed, Wallace turned instead to the great, beckoning constellation of small and large islands that stretched eastward for almost two thousand miles between Singapore and New Guinea. This region, corresponding roughly to what nowadays is Indonesia, was known then as the Malay Archipelago.

The biggest of the islands was Borneo, directly east, with Java just south of it; beyond those two lay Bali, Lombok, Celebes, Ambon, Flores, Timor, Komodo, Ceram, and thousands of others, including a little cluster at the eastern extreme, Aru, famous for its bird-of-paradise populations. Wallace had started learning the Malay language, which made remote travel throughout the archipelago more feasible. He was aware, from Darwin's *Journal* among other sources, that islands could be extraordinarily rich in endemic species. If they were rich in species, and bounded every which way by saltwater barriers to dispersal, they would also exhibit striking biogeographical patterns—that is, a richness of suggestive information. Wallace caught a boat to Borneo, where he could expect to be welcomed—thanks to a chance encounter—at the highest official level.

Along the north coast of Borneo stretched a strange private kingdom known as Sarawak, ruled by a buccaneering Englishman named James Brooke, the so-called White Rajah. Brooke had crossed paths with Wallace back in England, evidently liked him, and offered him hospitality if he ever got to Sarawak. Wallace turned up and, with Brooke's blessing, settled himself in a small house near the mouth of the Sarawak River. Brooke's own compound was farther upstream, so Wallace was alone again, except for a Malay cook. By this time it was early 1855, well into the wet season, with daily rains that made collecting bad or impossible. When monsoonal downpours wash a tropical forest, the butterflies and the birds hunker invisibly, the beetles crawl somewhere snug, and a person can barely see, let alone walk a trail, wave a net, or place a delicate creature into a dry jar. Wallace sat in his house, thwarted from doing fieldwork, and took the occasion to write another paper. This one was more ambitious than his account of Amazon monkeys, let alone any of the descriptive little reports he'd published on insects and fishes. He titled it "On the Law Which Has Regulated the Introduction of New Species."

He was groping toward a theory of transmutation. But he wasn't quite clear in his own mind as to how far his thinking had gotten him. There's evidence that he had already begun making notes for a book on the subject, which he planned to call *On the Organic Law of Change*. The book project was premature, and Wallace seems to have realized that much, or at least intuited it, confining himself for the meantime to producing this brief paper, which he would later refer to as "only the announcement of the theory." In truth, it wasn't even an announcement, because he didn't yet have a theory to announce. More accurately, the Sarawak paper was a hint about the existence of a phenomenon, transmutation, for which an explanatory theory would be necessary. The theory itself still eluded him. Unlike Darwin, Wallace was eager to put his exciting ideas into print, even if they were still amorphous.

Compounding the confusion, he chose some blurry terminology, which would eventually cause certain readers (including Darwin) to mistake his meaning. In the paper's title, for starters, his reference to the "introduction" of new species was deceptively bland and ambiguous. It seemed to imply a divine introducer. He also wrote about the "creation" of new species as modifications of earlier types. He used the term "antitype" for those precursors, suggesting a contrast or opposition (*anti-*), though what he probably meant was merely *ante-*type, a type that preceded. The "law" he had formulated was really just a generalized descriptive statement. It specified no mechanism of cause and effect.

Wallace nevertheless made big claims for this law. The puzzling patterns of biogeography, as well as the record of extinct species in geological strata, "are all explained and illustrated by it," he bragged. Casting light on biogeography was a valuable service, he noted, because so many odd facts had piled up since the time of Linnaeus, and no one had rationally accounted for them. For instance: "Such phenomena as are exhibited by the Galapagos Islands, which contain little groups of plants and animals peculiar to themselves, but most nearly allied to those of South America, have not hitherto received any, even a conjectural explanation." That was a gentle poke at Darwin, the most famous of Galápagos travelers, whose *Journal* had offered observations but no theory. Unaware that Darwin had held anything back, oblivious that Darwin's explanation was in progress but long overdue, Wallace couldn't have guessed that he was jabbing such a sensitive spot.

Wallace's law also solved—according to Wallace, anyway—the problem of systematic classification, by supplying a natural basis for grouping species into categories. It integrated Charles Lyell's vision of gradual geological changes into an understanding of trends in the fossil record. And it made sense of rudimentary organs. In politely phrased sentences, the young man was crowing loudly, an upstart naturalist on the far side of the world presuming to deliver a major revelation on the history of life to his better-educated, better-connected elders. He stated his law twice, once near the paper's start, again near the end, both times in italics so it couldn't be missed: "*Every species has come into existence coincident both in space and time with a pre-existing closely allied species.*"

What exactly did he mean by "come into existence"? What was implied by that coy word "coincident"? Should a "closely allied species" be understood as a genealogically related species? Wallace didn't say. When species do come into existence, is it by materialistic transmutation or divine creation? The answer to that question, clear in Wallace's head, was less clear on the page. If he did mean transmutation, what's the mechanism? He didn't yet know.

Never mind. It was a start. He mailed the new manuscript to Samuel Stevens, who would pass it along to an editor at the *Annals and Magazine of Natural History*, the same journal that had been hospitable to his Amazon field reports. Then he went back to other work. On a good day Wallace could collect, and in the evening there would be beetles and butterflies to pin. On a bad day he could read and think. In northern Borneo, the rain was still coming down.

22

Wallace knew Mr. Darwin, at this point, only from a distance. If they had met at all, it was passingly, at the British Museum, in the months before Wallace sailed for Singapore, an encounter that hadn't seemed important to either of them. So far as Darwin was concerned, Wallace was just another callow young traveler and commercial collector; his *Narrative of*

Travels was too weak on factual detail to impress serious naturalists in Darwin's circle. So far as Wallace was concerned, Darwin was merely the author of that fine *Beagle* journal, a robust but conventional book of natural history and exploration. He had no reason to suspect that Darwin was a transmutationist, like himself, and he wasn't interested in barnacle taxonomy. Having made his own four-year expedition into remote areas of tropical wilderness, under circumstances more difficult than anything Darwin had endured, he may have lost his sense of awe toward the man he had once put in a class with Alexander von Humboldt. Nothing came of the encounter . . . for a while.

Darwin's seemingly endless work on barnacles finally did end, in early autumn of 1854, about the time Wallace decided to leave Singapore for Borneo. Darwin noted in his diary, with a groan of frustration, that the project had cost him nearly eight years. The last of his four Cirripedia volumes wouldn't be published until weeks later, but on September 9 he finished packing up his specimens. He'd had enough of the squinty dissecting, the laborious sketching, the microscopic penises and fluttery legs. He was eager to move on. That very day, according to another diary note, he "began sorting notes for Species Theory."

It was back at the center of his desk. He'd had sixteen years to think about the transmutation of species, to refine his idea of natural selection, to mine the biological literature for relevant facts, to ponder his own data on variation and adaptation in the wild, to hone the arguments he had outlined in 1842 and drafted in 1844. Meantime he had also fathered nine children, buried two of them, and sent his oldest son off to boarding school. He had published eight books (not counting

the edited volumes of *Beagle* zoology), seven of them technical and one a popular travel narrative. He had made himself
an expert on the classification of a difficult group of animals,
and his expertise had been certified by a major award. No one
had a right, he had once worried, "to examine the question of
species who hasn't minutely described many," and now he had
earned that right. So was it time to *publish* his theory? No, not
yet. He still wasn't ready.

Instead he launched a further program of empirical
research to fill some of the gaps in his trove of evidence. He
became an experimentalist, cluttering his house and grounds
with simple but astute science projects that often smelled bad
but supplied useful data. He worked his network of far-flung
contacts for answers to obscure questions. He started keeping
pigeons. Throughout the next two years he concerned himself
largely with the anatomy and development of domestic animals, plant hybridization, plant fertilization, patterns of
species diversity reflected in plant classification, and the
capacity of plants for transoceanic travel.

How long, Darwin wondered, can a cabbage seed be soaked
in salt water and still germinate? How long for a radish seed?
A carrot seed? A kidney bean? A pea? He was curious about
what he called "accidental means" of plant species dispersal,
which might entail a seed, a pod, or a seed-bearing stem floating passively across a wide stretch of sea. So he tested the salt-
resistant viability of a whole list of vegetables and other
plants: rhubarb, asparagus, celery, cress, capsicum, furze, bar-
ley, and more. He mixed up a salt solution resembling sea
water, poured it into bottles, dropped in seeds as though they
had fallen on the ocean, and left them to ride the brine, or to
sink and marinate, for measured stretches of days. From this

set of experiments he learned several things. He learned that asparagus could float for twenty-three days if it were green and succulent, or as much as eighty-five days if dried first, and that the seeds would remain viable. He learned that cabbage and radish seeds became putrid and stinky "in a quite extraordinary degree," but that the radishes would still germinate after forty-two days of soaking, while the cabbages wouldn't. He found that cress seeds put out "a wonderful quantity of mucus," but that they too would sprout after forty-two days' immersion. Given the average rate of ocean currents, Darwin calculated, forty-two days was long enough for a floating seed or a pod to travel 1,300 miles. Most of the other species he tested were able to produce at least some germinations after twenty-eight days. The conclusion that Darwin drew from these experiments involved biogeography: Plants were certainly capable of crossing oceans. It didn't take an ancient land bridge that had sunk beneath the sea (as some of his colleagues imagined), and it didn't take an act of God, to explain how vegetation might appear on a new volcanic island.

Floating seeds weren't the only means for a plant to colonize across water. There were winged seeds, and tiny seeds with parachute rigging like those of a dandelion, which could travel on the wind. Another possibility was transport by a bird—a living bird, or even a dead one. Seeds might stick to the muddy legs of a heron or an egret and be rinsed off in a new location. His young son Francis, now eight years old, made a boy's gory-minded suggestion about dead birds—such as those that fall victim to a hawk, or to lightning, or maybe to apoplexy—and Darwin pounced on it. He floated a dead pigeon in salt water for thirty days. Seeds from the pigeon's crop germinated nicely.

Another bird-related experiment went to the question whether small animals, such as snails, might hitchhike from one place to another. Darwin cut off a pair of duck's feet and suspended them in an aquarium full of freshwater snails. If a duck was asleep on the water's surface, dangle-footed and oblivious, how many snails might climb aboard? Would they cling tightly when the duck flew away? Darwin waved his duck feet through the air. How long would the snails stay alive out of water? He let the snails languish overnight. His results suggested that freshwater snails could catch and survive a ride of six hundred miles.

He wondered too about lizard eggs. Would they float on seawater? For how long? Having floated for a month or so, would they still hatch? He offered to pay schoolboys a shilling for every half-dozen lizard eggs they could find; snake eggs welcome, too. He'd float them in his cellar. It was all relevant, just as the seed-brining experiments were relevant, because transmutation implied the necessity of natural dispersal. If there was no special creation, there was no special delivery. Biogeography, from a transmutationist perspective, reflected the fact that species had arisen one from another, adapted, and traveled. Darwin needed to prove, among other things, how well plants and animals could get around.

He also wanted measurements of different varieties of domestic animal, especially fetuses and juveniles, in order to see how their differentiation in form during growth and development might echo evolutionary divergence from common ancestors. He asked friends to keep his morbid interests in mind when any of their pets or livestock died. In one letter, to W. D. Fox, he begged for a week-old chicken and a nestling pigeon, from which he meant to make skeletons. He mentioned passingly to Fox that he'd already begun comparing wild and tame ducks. When he got live birds, he killed them with chloroform or ether, boiled the carcasses to soften them, and then stripped off the flesh, a smelly process that often made him vomit—and not just him, with his delicate stomach, but also Parlsow, the all-purpose butler. So he outsourced that phase of the work. About mammals, he reported cheerily: "I have puppies of Bull-dogs & Greyhound in salt." And he had commissioned someone to make careful measurements of young colts, both racehorses and cart horses. Whenever possible he wanted data from standard-aged juvenile forms, so that comparisons were valid; for birds, he tried to get them seven days after hatching. But juveniles of some species and some breeds weren't always easy to find. Did anyone know, he asked, how to lay hands on a seven-day-old wild duck?

He solved the availability problem for pigeons by setting up his own breeding operation in a backyard aviary. He was interested in the fancy breeds, the pouters and fantails and tumblers and English carriers and others, whose extravagant shapes and behaviors reflected hundreds of years' selective breeding by proud, obsessive pigeon fanciers. "The fancy," as it was called, wasn't an expensive avocation and some of those fanciers were workingmen, who coddled and bred their birds in London rooftop coops and talked the subtleties of coloring, beak shape, carunculated eyes, and feathery decor in their local pubs. Having begun with the cold detachment of an experimentalist, Darwin found himself charmed by the pigeons and amused by this subculture surrounding them. He studied breeders' manuals, corresponded with experts, read *Poultry Chronicle*, went on pigeon-shopping excursions in London, even joined two of the city's pigeon-fancier clubs. At the height of his own fancy, he had sixteen different breeds. "I am getting on splendidly with my pigeons," he told his son William, the one at boarding school. He'd just added some trumpeters, nuns, and turbits, plus a small pair of German pouters, given to him by a brewer pal in London. In the summer, Darwin confided to Willy, he looked forward to flying his tumblers. So much for the cold heart and the sharp knife of science.

Late in 1855 he drafted a form letter, a generalized request he intended sending to overseas contacts and acquaintances. It was phrased like a classified ad under the WANTED heading. "Skins," it began: "Any domestic breed or race, of Poultry, Pigeons, Rabbits, Cats, & even dogs, if not too large, which has been bred for many generations in any little visited region, would be of great value." He was asking a sizable favor: Please ship me specimens. In addition to the skin with its feathers or fur, he wanted a humerus and a femur and as much as possible of the cranium, preferably all still connected by sinew. The part about "many generations" in a "little visited region" was important for his studies of how individuals within a given population vary. Darwin now recognized that this crucial phenomenon, variation, occurs constantly in wild species as well as in domestic stock—but what *causes* it? Huge question. He didn't know. One possibility, he thought, was differences in external circumstance. So he hoped to see how domestic breeds might vary when raised in exotic locales such as Persia, Jamaica, or Tunisia. He would happily reimburse the costs of skinning and shipping.

He made a list of the men to whom this request went. It included such figures as Rajah James Brooke (in Sarawak), Sir John C. Bowring (governor of Hong Kong), Sir Robert

Schomburgk (an explorer of Guiana, and then British consul in Santo Domingo), the botanist G. H. K. Thwaites (in Ceylon), E. L. Layard (a museum curator in Capetown), and Edward Blyth (another curator, in Calcutta). Blyth would become one of Darwin's most helpful and prolix respondents. Halfway down the list appeared an inconspicuous name, "R. Wallace," unaccompanied by any geographical notation. Darwin evidently had a mailing address of some sort for Alfred Russel Wallace—possibly the one in Sarawak, Wallace's temporary base—but at this point he couldn't have guessed exactly where, within the vast Malay Archipelago, Mr. Wallace might actually be. And they barely knew each other. Darwin was just tossing a penny into a well.

23

Wallace's paper from Sarawak, about the "law" regulating the "introduction" of new species, was published in September 1855. It created no sensation, but it did generate some murmurs. Wallace's agent, Samuel Stevens, told Wallace about several London naturalists who had groused that he should stop theorizing and stick to collecting facts. Charles Lyell, on the other hand, found the paper intriguing. Out in Calcutta, Edward Blyth got his copy of the *Annals and Magazine of Natural History* and reacted similarly. In one of his long letters to Darwin, near the end of the year, Blyth asked: "What think you of Wallace's paper in the *Ann. M. N. H.?*" His own answer: "Good! Upon the whole!" Darwin's opinion was different. He read the paper around that time and made some notes for his own memory, as he routinely did with his eclectic research reading. That was Darwin's way, methodical and thorough; he chewed through huge amounts of material, swallowed the good bits, spit out the rotten stuff and the husks. Wallace's paper tasted like husk.

It discussed geographical distribution, Darwin recorded, but offered "nothing very new." It used the simile of a branching tree ("my simile," in Darwin's jealous view) to represent affinities and diversity in nature. It mentioned rudimentary organs, though to what point? And the Galápagos comment —about how those peculiar creatures and curious patterns had never received "even a conjectural explanation"— didn't pass unnoticed. Darwin may even have winced, knowing it was true. He *hadn't* risked any explanation in the *Journal*, but . . . give a man time. Well, all right, he'd *had* time. Still, not enough. And what did Mr. Wallace know of the complex considerations? Rather than arguing the point in his mind, or rising to this small provocation as a challenge, Darwin dismissed Wallace's whole effort. He saw no real explanatory value to the "law" of juxtapositions and he heard nothing in the vague language except a rehash of old-fashioned natural theology. Now if Wallace had scratched the word "creation" and spoken instead about "generation" of new species, Darwin told himself, he could agree with the paper. So far as it went. But Wallace hadn't used any such word. "It seems all creation with him," Darwin judged, and went back to his pigeons.

He sent off his letters to Thwaites, Layard, and those others on the list, including "R. Wallace." I would be most grateful, he told them, for any skins of chickens, pigeons, rabbits, or ducks.

24

After a year in Sarawak, Alfred Wallace shifted onward in the Malay Archipelago to find new hunting grounds, beyond the range of previous British travelers and collectors. He caught a Chinese-owned schooner that stopped briefly at Bali and then deposited him on Lombok, a small island just thirty miles further east. Wallace stayed on Lombok for two months, shooting birds and observing the local culture while waiting for another boat that would take him to Macassar, a port on the bigger island of Celebes. Lombok is where he first encountered the sulfur-crested cockatoo, a gorgeous but noisy bird not found on Bali or any of the other islands westward. He also noticed the rainbow bee-eater, another pretty species common in Australia. Wallace would eventually realize from these signals and others that, just in bouncing from Bali to Lombok, across a narrow but deep strait, he had moved from one biogeographical zone into another. He was now in the realm of Australian fauna. That seemed odd. Why should there be such well-demarcated zones?

From Lombok, he sent off a crate of specimens, to Stevens in London by way of Singapore, containing more than three hundred bird skins. Most of those, including as many cockatoos as he'd been able to kill, were intended for sale. The crate also contained something so ordinary that, coming from a commercial collector of biological exotica, it must have seemed peculiar: a local variant of the barnyard duck. Wallace's note to the agent explained: "The domestic duck var. is for Mr. Darwin." Please forward.

It's hard to say whether that duck ever reached Darwin. If so, he was presumably grateful but not surprised. He had come to expect a high degree of generous cooperation from the people (especially those below him in social status) he called on for research assistance. Around the same time, Wallace wrote to him directly. Sent from Celebes, traveling the slow mail routes of the day, this letter took six months to reach Down House. Like Darwin's first note to Wallace, it hasn't been preserved in the huge archive of Darwin correspondence; its existence and contents can only be inferred from the reply it evoked. "By your letter, & even still more by your paper in Annals," Darwin wrote Wallace on May 1, 1857, "I can plainly see that we have thought much alike." Choosing his phrases with some delicacy, he added that "to a certain extent" they had reached "similar conclusions." Furthermore, Darwin said, he endorsed "almost every word" of Wallace's paper and considered it rare for two theorists to agree so closely. Given Darwin's cold dismissal of the "law" paper in his reading notes—"nothing very new"—this was spreading the butter a bit thickly.

But something had changed. Wallace's lost letter may have contained a declaration of transmutationist views, and maybe also a boast that his paper was just the first step toward an explanatory theory. Such news would have put Darwin on guard. In any case Darwin knew that Wallace, consciously or not, was noodling along the edges of transmutationism. How consciously, how fruitfully? Those were separate questions, which Darwin doesn't seem to have asked. Seeing the younger man as a diligent but unsophisticated field naturalist and a possible but unlikely competitor, he was happy to share facts and vague musings, and careful to get more than he gave. Several other factors, in the meantime, had affected Darwin's state of mind.

First, he had grown more impatient to reveal his big secret. He confessed to his old chum Fox, a non-scientist, that his present work involved the question whether species are immutable or not. (He knew his own answer to that but pretended to be uncertain.) He hoped to produce a book on the subject within a few years, he told Fox. With several scientific colleagues he went further, admitting his conviction that species do transmutate and outlining his theory. Joseph Hooker was already in the know, having by now read Darwin's unpublished 1844 essay, but early in 1856 Darwin revealed his thinking to Charles Lyell and two or three others, including T. H. Huxley, the brilliant anatomist and popular lecturer who taught natural history in London. Huxley and Hooker and their wives, along with one other scientist, paid the Darwins a weekend visit at Down House in April, during which the host spilled his beans about natural selection. Skeptical of religion, argumentative by nature, Huxley greeted Darwin's wild idea with wild enthusiasm but managed to keep the secret when he went back to London. Lyell and his wife also visited the Darwins for a few days that month, and on the morning of April 16 the two men had a quiet talk. Darwin laid out his heretical, ingenious theory. He must have gulped hard first, given that Lyell had lambasted Lamarckian transmutationism in *Principles of Geology*. Lyell's reaction was strong but complicated, reflecting both his intellectual courage and his attunement to the imperatives of a scientific career. He didn't accept Darwin's notion, not yet. He did recognize its power and importance. In a private journal of his own, devoted to the species question, Lyell faithfully summarized that day's discussion. Remembering what Wallace called the "law" of closely allied species appearing adjacent in space and time, Lyell acknowledged that Darwin's theory of natural selection seemed to explain it. He sensed that these two hounds were running the same hare.

The second new factor in Darwin's mental mix was that Lyell gave him some pointed advice: Publish. Enough delay, enough caution, enough perfectionism. Go to press. "I wish you would publish some small fragment of your data," Sir Charles wrote shortly after the visit; "*pigeons* if you please & so out with the theory & let it take date—& be cited—& understood." He was still a creationist himself, but also a loyal friend. Vicariously, on Darwin's behalf, he felt the urgency to announce this great discovery—or anyway, this dramatic idea—and claim credit.

Darwin promised to consider Lyell's suggestion. But he was reluctant and, as he admitted, confused. To come "out with the theory"—that was easier said than done. How could he do justice, in a hasty synopsis, to such a provocative and complicated set of facts, inferences, and concepts? How could he make the theory persuasive without presenting all his evidence? How could he answer preemptively all the objections he expected? And what was the rush? He felt strung between scientific ideals and scientific ambition. "I rather hate the idea of writing for priority," he told Lyell, "yet I certainly shd. be vexed if any one were to publish my doctrines before me." That sentence captures it: He hated the idea of writing for priority, but dammit he did want priority.

A week later he wrote to Hooker, his closest friend, with whom he could be even more candid. "I had a good talk with Lyell about my species work, & he urges me strongly to publish something." A journal article covering part of the subject, for instance. Or I might do a very thin volume, Darwin said, except that it's "dreadfully unphilosophical" to publish such a thing without detailed factual support and references. He didn't want his work to look glib and racy, like *Vestiges of the Natural History of Creation.* "But Lyell seemed to think I might do this, at the suggestion of friends," he told Hooker, "& on the ground which I might state that I had been at work for 18 years, & yet could not publish for several years." The special pleading had begun.

He "could not publish" for several years. Why? Because he was cautious and methodical and hadn't yet set himself to the writing. Because he had chosen to proceed slowly. Now he was of two minds: He wanted to put something on record, for the sake of claiming priority, and he wanted to delay publication, for the sake of better preparing his case. He wanted to do it for his own peace of mind, but he preferred to say that he'd been persuaded by friends. Over the full sweep of his life, Charles Darwin was a man of great integrity, great goodness, deep generosity, and considerable courage; this episode puts his strengths in relief by showing him in some of his weakest, least forthright moments.

Hooker's reply, which also hasn't survived, argued against Lyell's suggestion of a journal article, though not necessarily against a "preliminary essay" in the form of a separate volume. A journal article, even in those days, implied some level of institutional vetting. But a little book, published privately at the author's expense, wouldn't implicate anyone except the author in its wild-eyed ideas. It wouldn't require editorial review or full citation of evidentiary sources. On the other hand, Hooker warned, releasing a slim volume now might undermine the impact of the big book that Darwin intended to publish eventually.

So the two trusted advisers offered conflicting counsel, and Darwin himself was flummoxed. He did begin drafting a short version—call it another essay, an article, whatever—but he soon grew frustrated at the effort to select and reduce so severely. By late summer of 1856 he had brushed off Lyell's advice and changed his approach, writing chapter by chapter on a scale that would eventually yield several fat, exhaustive tomes commensurate to Lyell's own three-volume *Principles of Geology*. He failed to appreciate (or decided subliminally to ignore the warning signs) that young Wallace was following the same intellectual route at a pace not braked by caution.

Toward the end of the year, Darwin wrote again to Lyell: "I am working very steadily at my big Book;—I have found it quite impossible to publish any preliminary essay or sketch; but am doing my work as complete as my present materials allow, without waiting to perfect them. And this much acceleration I owe to you." It wasn't acceleration enough.

25

Darwin and Wallace were now tenuously in touch but communicating at cross purposes through the international mails. Did the duck sent from Lombok ever make it to Darwin's dissecting bench? My guess is no, because the bird goes unmentioned after Wallace's cover note. Maybe the whole shipment of specimens went astray. Maybe it reached Samuel Stevens in a state of unpresentable rot. Anyway, no trace of a thank you appears in Darwin's otherwise unctuous letter of May 1, 1857, in which he conveyed his compliments on Wallace's "law" paper.

There's another odd comment in that letter, showing Darwin's sensitivity about how long he had delayed. After noting the similarity of their views, and the rarity of such concord between two theorizing naturalists, he stroked a dash on the page, as though clearing his throat. Then he wrote: "This summer will make the 20th year (!) since I opened my first notebook, on the question how & in what way do species & varieties differ from each other." At last, Darwin intimated, he had found the answer. Anyway he'd found *an* answer, a distinct and tangible idea. Others would judge whether it was right or wrong. He couldn't possibly explain this idea in a mere letter, he told Wallace, too complicated. "I am now preparing my work for publication, but I find the subject so very large, that though I have written many chapters, I do not suppose I shall go to press for two years." He was wheedling for time and consideration.

Although he still didn't take Wallace quite seriously—not seriously enough—he felt mildly wary. With its histrionic exclamation point, Darwin's remark was an assertion of his own interests, precedence, and claims. A male dog makes the same sort of assertion, raising his leg to mark a tree. Wallace's nose must have been off, because he didn't get the hint.

His Abominable Volume

1858–1859

26

On or about June 18, 1858, another mailing from Alfred Wallace arrived at Darwin's front door. It came, like the others, from somewhere in the Malay Archipelago. It had been four months in transit on a series of boats. This envelope was bulkier than usual, containing a manuscript as well as a letter. Darwin opened it. Scanning the letter, reading the enclosure, he felt a nauseating surge of emotions that began with surprise and swelled quickly toward despair. His big book at this point was still a work in progress, two-thirds written and growing more unwieldy every day. Meanwhile his young pen pal, Wallace, had independently conceived the idea of evolution by natural selection.

Wallace's manuscript was titled "On the Tendency of Varieties to Depart Indefinitely from the Original Type." It comprised about twenty pages of lucid, easy prose, written out in the author's hand. Its cardinal point, signaled in the title, was that the difference between species (as a category) and variety (as a category) is merely a difference of degree. That is, the amount of variation seen between varieties within a species is not inherently limited; rather, those increments can accumulate boundlessly until a variety splits away, becoming a distinct species unto itself. The manuscript posited "a general principle in nature" causing many varieties to do exactly that. And they don't just *split* from the parent species, Wallace asserted; they compete against it, sometimes outlive it, and eventually give rise to still other varieties differing more and more from the original type. Wallace, unlike Darwin, had coined no name for this "general principle." But his manuscript built a case for it with logic very similar to Darwin's own.

"The life of wild animals," Darwin read, "is a struggle for existence" in which "the weakest & least perfectly organized must always succumb." That struggle is driven by the pressure of inherent population growth rates, yielding many more newborn individuals than can be supported by available food and habitat. Without mentioning Malthus by name, the manuscript gave a deft summary of Malthusian arithmetic. It noted that "variations from the typical form of a species" occur commonly among wild animals (as Wallace, the commercial collector, had often seen), and that most such variations "would affect, either favourably or adversely, the powers of prolonging existence." For instance: "An antelope with shorter or weaker legs must necessarily suffer more from the attacks of the feline carnivora." Lions would eat the slow ones. A passenger pigeon with less powerful wings would have trouble traveling widely to find food. Starvation and competition would eliminate the poor flyers. On the positive side, a giraffe with an especially long neck would have access to high leaves that the others couldn't reach. During a famine it might sustain itself with that extra resource, while short-necked giraffes died away. Those creatures "best adapted" as a result of such small differences would eat better and defend themselves better, survive better and reproduce more abundantly, establishing sizable populations while less fortunate creatures lost the struggle and disappeared. The result would be "continued divergence" over long stretches of time, with "successive variations departing further and further from the original type." Wallace's manuscript ended with a flourish, suggesting that "all the phenomena presented by organized beings, their extinction & succession in past ages, & all the extraordinary modifications of form, instinct, and habits which they exhibit" are accountable to that nameless "general principle in nature."

A big claim. The cover letter was more modest. Here's a hypothesis I've hit upon to explain the origin of species, Wallace said. He hoped it might seem as new to Mr. Darwin as it had to him when the notion first struck.

It didn't.

27

The manuscript was datelined: "Ternate, February 1858." Wallace had mailed it from a tiny volcanic island in the northern Moluccas. As the story goes, he got his inspiration during an attack of malarial fever, while forced to lie bedridden suffering alternate cold and hot fits and unable to do anything but think. One thing he thought about, as he'd been doing for years, was how species come into existence. Having seen such a spectrum of variation in the wild, having charted the suspicious distribution of closely allied species, Wallace had become ever more persuaded of the reality of transmutation.

But what was the causal mechanism? During his bout of fever, he happened to remember Malthus, whom he'd read more than a dozen years earlier. He recalled the geometric rates of population increase, the slower increases in available food, the consequent "checks" to human population growth. Suddenly it occurred to Wallace, just as it had to Darwin, that such checks also regulate animal populations in the wild. Pondering all that adversity and mortality, he asked himself why some individuals survive while so many others die. "And the answer," as he recollected long afterward, was that "on the whole the best fitted live." Accidental variation plus the imperatives of struggle result in differential survival; differential survival leads to adaptation; divergent adaptation over vast stretches of time leads to fleet antelopes, strong-winged pigeons, and tall giraffes. Bingo. "The more I thought over it the more I became convinced that I had at length found the long-sought-for law of nature that solved the problem of the origin of species."

When the fever broke, he got up and scribbled some notes. Within a few days he had written his manuscript and sent it to Darwin by the mail steamer that stopped at Ternate.

Why had Wallace chosen Charles Darwin, of all people, to receive an outpouring of his febrile idea? It wasn't because Wallace recognized Darwin as a fellow transmutationist. The older man, in his published writings and their few exchanged letters, had been too coy to give that much away. As far as Wallace knew, Mr. Darwin was merely a conscientious naturalist of the traditional sort, whose interests ran to biogeography, barnacles, and variation in poultry. But Wallace, thrilled with what he'd hit upon, eager to announce it, had to send the manuscript to *somebody*, and his options were limited. He'd already heard through Samuel Stevens of the derogatory mutters, in London, about his ventures into theory. The old boys at home thought he should stick to gathering salable beetles. He might well have ignored those dour signals and mailed his paper to Stevens anyway, for forwarding to the *Annals*, as he'd done with the earlier writings. But that didn't seem wise, not this time; the stakes were too large, the concept too incendiary. Or maybe he was simply aiming higher. Who else did he know? Wallace was isolated out there in the islands, and not just by miles and water. His lack of scientific credentials, educational polish, and social position left him feeling marginal. He had grown discouraged by a sense that his "law" paper came and went quietly, attracting almost no notice. He had even complained about that in a letter to Darwin. Darwin had responded, as a kindly aside, that it wasn't quite so—that Darwin's friend Lyell, for one, had found the "law" paper intriguing.

Lyell had? Sir Charles Lyell, Britain's preeminent geologist? This was a delicious bit of flattery for Wallace's modest ego. Now, half a year later, Wallace was hoping to play the Lyell connection. If the enclosed manuscript on species seems sufficiently important, he asked Darwin, would you please pass it along to Sir Charles?

28

Darwin felt crushed. He had only himself to blame. His dilatoriness, his perfectionism, his big mouth. Suddenly he was trapped, flattened, between the demands of honor and the claims of self-interest. He howled with pain. "Your words have come true with a vengeance," he wrote Lyell, "that I shd. be forestalled." Enclosed is a manuscript that Wallace asks me to send you, said Darwin. It's well worth reading. It's also, he added glumly, the closest thing to a précis of my own theory. (In the panic of the moment, he was overlooking a significant difference: Wallace focused on competition between varieties, not between individuals—that is, selection of one group versus another, not selection of individuals within a group.) "I never saw a more striking coincidence," Darwin moaned. Even some of the phrases Wallace used, such as the "struggle for existence," echoed what Darwin had already written into the draft of his big book. Wallace hadn't asked him to help get the manuscript published, Darwin noted, only to share it with Lyell; but of course Darwin would write Wallace immediately and offer to send it to any journal. "So all my originality, whatever it may amount to," he whined, "will be smashed."

Lyell, always a steady head, advised him to calm down. Maybe there was an alternative solution, less drastic than all-or-none priority. Joseph Hooker, also a sensible friend as well as a faithful one, was brought into the discussion. As days passed and letters flew back and forth, though, Darwin's attention became split between the Wallace surprise and some family concerns not conducive to calm.

A wave of sickness hit the village and the household. His eldest surviving daughter, Etty, caught a sore throat that turned out to be diphtheria, a scary and relatively unknown disease in those days, which was running at epidemic levels across Britain. By the time Etty improved, there was another disease to fear: scarlet fever, which had broken out locally. Three children in the village died, others were hovering in danger, and on June 23 it hit baby Charles, the youngest Darwin.

This namesake child is a mysterious figure, about whom evidence is scarce and scholars disagree. Born when Emma was forty-eight, christened Charles Waring Darwin, by the age of nineteen months he was a toddler who hadn't toddled. Small for his age, he didn't walk or talk. He had a sweet, tranquil disposition but seldom laughed, seldom cried, and made weird faces when he was excited. Evidently he had some sort of physical and mental debility, though it's hard to say what. According to Etty's later testimony, her littlest brother was born "without its full share of intelligence." Of the two best and most thorough Darwin biographies, Janet Browne's and the Desmond-and-Moore, the latter describes baby Charles as "severely retarded," the former says that he "may have been slightly retarded," possibly because of mercury poisoning from ill-conceived Victorian medicines. Randal Keynes, Darwin's great-great-grandson, argues persuasively that Charles Waring suffered from Down Syndrome—that is, physical impairments resulting from an extra copy of the twenty-first chromosome. It was a perplexing condition at the time, not even partly clarified until Dr. John Langdon Down (no connection to Downe village or Down House) identified it eight years later. Whatever his trouble, baby Charles was loved by Darwin and Emma with a pitying tenderness that probably included some sense of burden and regret, making their feelings all the more complicated when he died of his fever on June 28.

The end was ugly and hard, as it had been with Annie. Otherwise that death and this one were almost as different as two child-losses could be. "It was the most blessed relief," Darwin told Hooker, "to see his poor little innocent face resume its sweet expression in the sleep of death." Etty later described her parents' reaction more bluntly, recalling that, "after their first sorrow, they could only feel thankful." Darwin wrote a short, private memoir of Charles Waring, in which he did his best to accentuate the positive, recalling "nice little bubbling noises" the child sometimes made, how "elegant" he looked crawling naked on the floor, his "placid & joyful" disposition.

In the meantime Darwin had heard back from Lyell with some thoughts about how the Wallace dilemma could be handled. What did Darwin have on paper, Lyell wondered, that might testify to his priority of discovery? Well, there was the manuscript essay of 1844, which Hooker had read; also a six-paragraph summary of the theory, which he'd sent last year to the botanist Asa Gray, his trusted correspondent at Harvard. These unpublished but witnessed writings were proof that he'd conceived the whole idea long ago, solitarily, and stolen nothing from Wallace. "I shd. be *extremely* glad *now* to publish a sketch of my general views in about a dozen pages or so," he told Lyell. "But I cannot persuade myself that I can do so honourably." He worried that receiving the Wallace manuscript—which he hadn't asked for, after all—put him in a bind. He would rather burn his own book-in-progress, he said, than be seen as behaving shabbily. But was it too late to publish a summary of his views and say he was doing so on the advice (two years earlier) of Lyell? He repeated: "If I could honourably publish. . . ." No, he couldn't persuade himself that it was okay; but implicitly he begged Lyell and Hooker to do the persuading.

Altogether, he was fuddled with anguish. He hated himself

for thinking about such stuff while his children were battling for their lives. "This is a trumpery letter," he ended, "influenced by trumpery feelings." But the feelings wouldn't go away.

Lyell and Hooker took their cue. Within days, serving him faithfully as friends, serving science by their lights, serving justice more dubiously, they cooked up an arrangement that rescued the situation—or at least, it rescued Darwin's interests. They certainly couldn't ignore Wallace's paper entirely and connive to see Darwin given credit alone; that would have been dishonorable, unprofessional, and scandalous when the truth came out. Instead they devised and sponsored a joint presentation of Wallace's manuscript and Darwin's unpublished work. This peculiar duet would occur at the next meeting of the Linnean Society, one of London's better scientific associations, of which Hooker, Lyell, and Darwin were all governing members. Darwin consented to the arrangement, sending Hooker his 1844 essay and the six-paragraph summary he'd written for Gray, along with another disclaimer: "I daresay all is too late. I hardly care about it." No wonder: The baby at that moment was still alive, but barely, fevering toward a crisis. Wallace, on the other hand, didn't consent to the joint reading (at least, not in advance); he couldn't, because no one consulted him. He was still doing fieldwork in the eastern islands, unreachable on short notice, far out of the loop. Nobody seems to have asked Lyell and Hooker: *Gentlemen, what's the all-fired hurry?* Nobody suggested that Darwin, having waited twenty years to publish, might reasonably wait another six months for Wallace's acquiescence. It was a done deal before anyone thought to quibble. The reason for hurry,

I think, was that Lyell, Hooker, and Darwin all felt some embarrassment about this high-handed bestowal of shared credit, and they knew that delay might bring complications.

So there was no delay. The insiders moved deftly and fast. The details were settled in a flurry of overnight letters between London and Downe. Hooker chose an excerpt from Darwin's 1844 essay and inserted that, along with the Gray summary and Wallace's manuscript, into an already full agenda for the Linnean Society meeting. These three statements were ordered alphabetically by author—Darwin's two, followed by Wallace's. On the evening of July 1, 1858, the Darwin-Wallace material and five other papers were read to an audience of about thirty people. Hooker and Lyell attended. By coincidence, so did Samuel Stevens, who may have wondered how this Wallace paper got to London without passing through his hands.

The two authors were absent. By hindsight you might view them as "conspicuously absent," although Wallace's non-presence wasn't notable at the time. He didn't belong to the Linnean Society. His voice was admitted like the *crawk* of an exotic parrot, interesting and indelicate. He spent July 1 at a place called Dorey, a trading village on the northwestern coast of New Guinea, five hundred miles east of Ternate. The wet season had struck again, collectable birds were scarce around Dorey, but for insects the hunting was excellent. He'd been doing especially well with beetles. He was unaware of the event in London.

Darwin, acutely aware, missed the Linnean meeting, too. He was home in Downe with a dead child and a bad case of ambivalence.

29

The most remarkable thing about Darwin-Wallace night at the Linnean Society is how little immediately came of it. No general discussion followed the reading of papers. No one stood up in response to what Darwin and Wallace proposed and said, *That's brilliant!* or *That's outrageous!* Tea was served, probably. There was some private chat. And then the Linnean fellows went home. The foundations of science had shifted beneath their feet but they didn't notice.

Why not? This is hard to know. Possibly it was because the excerpts from Darwin and the paper from Wallace focused on the circumstances and details of the mechanism, natural selection, not on its larger significance. The word "transmutation" wasn't mentioned by either author, let alone the word "evolution" (though Darwin did allude to "the origin of species"). In the ears of a careless listener, on a hot July night, during an overlong meeting, the Darwin-Wallace readings with their roundabout logic may have seemed to involve merely varieties and variation. Another reason that the audience missed the point may have been that those Linnean fellows generally weren't asking themselves the question—*How do species change, one into another?*—that Darwin and Wallace were answering.

Two months later, the society's *Journal of Proceedings* published Darwin's fragments and Wallace's manuscript, lumping them as though they were a single co-authored paper. In the editing process, someone supplied a slightly garbled portmanteau title: "On the Tendency of Species to Form Varieties; and On the Perpetuation of Varieties & Species by Natural Means of Selection." Printed, the three pieces carried more impact than they had in spoken delivery. At least a few scientists recognized that this was weighty stuff, for better or worse, though some only condescended to disparage it. The president of the Geological Society of Dublin declared to an audience, early the next year, that Darwin and Wallace's paper "would not be worthy of notice" if Lyell and Hooker hadn't acted as sponsors. "If it means what it says, it is a truism," according to this man; "if it means anything more, it is contrary to fact." Darwin heard about that criticism and savored it as "a taste of the future." He was right.

Other readers came across the Darwin-Wallace amalgam and were deeply affected. "I shall never forget the impression it made on me," one naturalist, young at the time, wrote later. Hooker included references to "the ingenious and original reasonings" of Darwin and Wallace in his forthcoming work on Tasmanian plants, and Asa Gray described the Darwin-Wallace theory to an elite science club at Harvard, causing a disagreeable buzz in the head of Louis Agassiz, the eminent professor of natural history. So there was a scattering of strong reactions, but no explosion of acclaim or alarm. Either the central idea shared by Darwin and Wallace was too shocking for immediate absorption, or else other circumstances, somehow, hadn't favored immediate uptake. Maybe the idea hadn't been clearly enough expressed, or not cogently enough supported with factual evidence—or maybe people just weren't paying attention. Anyway, it slipped past. When the president of the Linnean Society (who happened to be Darwin's old reptile identifier from the post-*Beagle* days, Thomas Bell) gave his annual address the following May, he offered a bland retrospect of the past year. It hadn't been enlivened, Bell said, by "any of those striking discoveries which at once revolutionize" a branch of science. Bell's comment is now famous for its obtuseness. But in a strict sense he was right. The joint Darwin-Wallace announcement didn't "at once" revolutionize biology. It was too elliptical and dry. Something more was necessary.

As for Darwin, he rebounded quickly from tragedy and despair. On Tuesday of the bad week, with little Charles dead and other household members in danger, he had told Hooker, "I am quite prostrated & can do nothing" beyond sending the excerpts. By the following Monday, he and Emma had packed off their healthy children to stay with her sister in Sussex, getting them clear of the house, while Darwin resumed his scientific correspondence. Work bolstered him if it didn't (in excess) sicken him. That hadn't changed.

Work was his opiate, and science was his religion. He wrote to Asa Gray about bumblebees. He wrote to one of his pigeon contacts, asking for a young turbit he could pickle and measure. Mostly his thoughts were on natural selection: how to salvage his discovery, what to do next. The arrival of Wallace's paper had jolted him into a new frame of mind. There could be no more procrastination. No more plodding perfectionism and encyclopedic amassment of facts. No more timidity. With nudging from Hooker, he seized the idea of writing a sleek "abstract" of his theory, short enough to be published in a journal. Not just a fragmentary piece of the logic and data, as Lyell had suggested ("*pigeons* if you please") two years earlier; no, this would be a small version of the full conceptual edifice. And he'd be sole author, of course, not entangled in a partnership with Wallace. Yes, that's it, he'd do an abstract. It could go to the Linnean Society *Journal,* for which Hooker played a guiding role. He set the big book aside and started fresh.

The new plan, and the Wallace scare, gave him new energy. During a July getaway to the Isle of Wight, complicated by transferring seven children plus servants into a seaside villa, Darwin wrote for several hours each day. His approach in these pages was slightly more brisk, more personal than in the half-written tome. He forced himself to focus on essentials. Choose the key points, hit them clearly, move on. Pick only the strongest and most vivid of his illustrative facts. He tried to construct a beguiling argument that might carry readers along, rather than erecting a mountain of data that would crush them into surrender. He wrote conversably, in the first-person singular and sometimes the first-person plural: "When we look to the individuals of the same variety or sub-variety of our older cultivated plants and animals. . . ." He even found himself enjoying the work, unusual for Darwin as a writer. He was telling a great tale, as he'd told a great tale in his *Beagle* book—and telling it not so much from his research portfolios as off the top of his head.

This felt liberating, at least initially. Having resolved to pour his theory into an abstract, he vowed to do nothing else until it was finished. He mentioned as much in a letter to his friend Fox. I'll send you a copy when it appears in the journal, he added jauntily.

He tried to be severely concise. That wasn't easy. Just a month earlier, panicked by Wallace, he had longed for the chance to publish a dozen-page summary of his views. Now he found himself unable to do justice to the subject in twelve pages or even in thirty or forty. There were too many angles. He knew too much. He'd need that many words just to cover variation in domestic species, only one topic among many he meant to include.

The abstract was growing, he warned Hooker at the end of the month. It would be longer than expected. Returning to Downe in mid-August, he continued to write. He drew on his twenty-year archive of data, his notebooks and reference books and correspondence and portfolios stuffed with bits of loose paper—but he drew selectively now, striking a balance between factual evidence and persuasive discourse. He omitted footnotes and full citations of his sources. He mentioned other researchers and informants only passingly. His little piece on variation among chickens, dogs, ducks, and pigeons grew into a chapter. He finished a chapter on the struggle for existence and another presenting his central idea, natural selection. In autumn he took a week's rest at a hydropathic spa—not Malvern, with its memories of Annie, but a closer place called Moor Park, in Surrey, with another flaky doctor treating his mystery illness. Then back to work. He wrote chapters on the laws of variation (as best he discerned them, which wasn't very well), hybridization, instinct, some objections that might be raised against his theory, and other topics. By the end of the year, concentrating fiercely, churning out pages like a hack, calibrating the line between too much and too little, he had produced about half of what would be a 500-page book. He was still calling it his "abstract," although soon it would be known as *On the Origin of Species*.

He relapsed into bad health in February 1859, suffering "the old severe vomiting," plus a swimming sensation in his head. "My abstract is the cause," he figured. Probably so. During another visit to Moor Park, where he could forget about species, almost, and amuse himself reading novels, playing billiards, flirting jovially with young women at dinner, he got some relief. He favored junky romantic novels with pretty heroines and happy endings, but he also enjoyed *Adam Bede*. He liked the billiards so much that he eventually bought himself a table. Back in the traces at Downe, he had only two chapters to write, followed by revising, and then he'd be "a comparatively free man," he told Fox.

Free of what? Free of the burden of secrecy? Free of the fear of preemption? Free of the duty to publish? Never mind, it was the casual comment of a tired man. Free of this damn book, anyway. But he was wise to qualify it: *comparatively* free. He would never escape the responsibilities and tensions that came with his great idea.

30

Wallace got news of the arrangement, by letter, when he returned to his base in Ternate. There wasn't one letter but two—from Darwin and from Hooker. Darwin's contained Hooker's as an enclosure, leaving Hooker to do the main explaining. Darwin was understandably abashed and tried to portray himself as a passive party swept along by events. (Later he would assure Wallace that "I had absolutely nothing whatever to do in leading Lyell & Hooker to what they thought a fair course of action," a claim that was weaselly at best and arguably untrue, given his strong hints and lamentations to both men. He would also misstate the dating of his own excerpts in the Darwin-Wallace package, telling Wallace that they'd been "written in 1839 now just 20 years ago!" In fact, they'd been written in 1844 and 1857.) Both letters to Wallace have been lost, but Darwin mentioned elsewhere that he considered Hooker's to be "perfect, quite clear & most courteous" in presenting the fait accompli.

How did Wallace react? Picture a lonely, long-suffering, self-educated man on a tropical island. Opening his mail, he learns suddenly that his malarial brainstorm of months earlier has yielded a theory that some of Britain's foremost scientists consider important—so important, even, as to be worth scuffling over. And he finds that, the scuffle having been settled without him, his allotted portion is a lesser half share of intellectual ownership and glory. His name is now famous, at least among the Linnean Society crowd; he has been recognized as a partner (a junior partner) with eminent, unimpeachable Mr. Darwin. His idea is launched, not just on the strength of his own arguments but with the authority of that unexpected co-discoverer. Well. Gracious sakes. It must have taken a few moments to sink in.

Maybe he spoke aloud to his dried beetles. There was no one around, in Dorey, with whom he could really share the news. He must have reread the letters, once, twice, tasting the words. Possibly he felt a twinge of resentment. So: *My* brilliant idea is now *ours*. Then, more wisely and cannily, Alfred Wallace decided that he was delighted.

The loss of sole credit was outweighed by another consideration: the honor, not to mention the practical advantages, of being welcomed as a colleague by these scientific insiders. To this honor Wallace responded gratefully, and with humility so dignified that in retrospect it seems almost smarmy. On October 6, probably just after receiving the two letters, he wrote to thank Hooker, endorsing the Linnean Society arrangement and declaring that it would have pained him if Mr. Darwin's "excess of generosity" had resulted in Wallace's paper being published alone. He was glad to know that Darwin had studied the same subject long and deeply. The more discussion of these facts and questions, the better. Too often a first discoverer gets all the credit, Wallace said, to the exclusion of another researcher who arrives at the same results independently. (He was right about that. Scientists, more competitive than many classes of people, generally race one another to make and announce discoveries, and the politics of priority were already intense in Darwin's day, even without grant-giving agencies and Nobel prizes. The adjudicated tie decreed by Hooker and Lyell went against standard scientific practice.) Notwithstanding its unusualness, their action had been "strictly just to both parties," Wallace claimed to believe, but if anything too favorable to him. He also wrote to Darwin, saying roughly the same thing.

That was his public response, anyway. Wallace's private response was more revealing. On the same day as his letter to Hooker, October 6, he wrote to his mother, Mary Wallace, also back in Britain. He was bursting. He wanted to share with her his excitement at having just received personal communications from two of England's most respected naturalists. He told her something (not much) about the circumstances—an essay he'd sent to Mr. Darwin, which was seen and admired by Dr. Hooker and Sir Charles Lyell, "who thought so highly of it that they immediately read it before the Linnean Society." Actually, *they* hadn't read it but only caused it to be read; that detail got lost in the telling to Mrs. Wallace. Anyway, who cared. It *was* read. And imagine: Darwin, Hooker, Lyell. In case his mother didn't get the implications, Alfred added: "This assures me the acquaintance and assistance of these eminent men on my return home." Look, Ma, he was saying proudly, I now have connections. And not only that—*chips* to play.

In his autobiography, almost fifty years later, long after

Darwin and Lyell were gone, Wallace would tell this part of the story a little differently. With a discreet editing stroke, he deleted his own hardheaded youthful opportunism. Quoting (but misquoting) the letter to his mother, he let himself say: "This insures me the acquaintance of these eminent men on my return home." By that time he had found and accepted his place, second place, in the pantheon of British evolutionary theorists. He hadn't had an easy life—no family money, no financial security, no institutional position, ever—but he was proud of his austere independence. That he had once coveted the "assistance," not just the "acquaintance," of powerful gentlemen was evidently discomfiting to recall.

31

In early May 1859, after just ten months of feverish work punctuated by short visits to the spa at Moor Park, Darwin finished a draft of his book. Chapter by chapter, it went to his private copyist to be made legible, and then to the printers in London. John Murray, whose publishing house had done Lyell's books and the successful second edition of Darwin's *Journal of Researches*, had agreed to publish it. Darwin began receiving proof sheets for correction at the end of the month and was appalled at how poorly his hasty prose read. The style was "incredibly bad, & most difficult to make clear & smooth." He'd never claimed to be much of a writer.

He revised heavily on the proofs, an expensive step in terms of additional typesetting costs, for which he offered to pay from his own pocket. His proposed title was deadly dull: *An Abstract of an Essay on the Origin of Species and Varieties through Natural Selection*. That reflected his lingering embarrassment at the absence of scholarly citations and the abridged selection of evidence; without full credit to sources and a full panoply of supporting data, Darwin felt, the work shouldn't be considered—or labeled—anything *but* an abstract. Fortunately, Murray and Lyell persuaded him otherwise. Murray was in business for profit, after all, not to perform a money-losing public service, and *An Abstract of an Essay* on whatever, species origins or Chinese bordellos, just didn't sound robustly commercial. In late September, when Darwin and Fox traded letters full of middle-aged grumping, Darwin reported that his health had been bad again but at least the book work was nearly done. Only an index to add, and then some revisions. "So much for my abominable volume," he wrote, "which has cost me so much labour that I almost hate it."

On October 1, 1859, Darwin finished correcting the proofs. By his calculation the abominable volume had taken thirteen months and ten days of concerted writing effort, not counting time off for rest, travel, billiards, and vomiting. In mid-October he warned Hooker what to expect—that he'd gone to great lengths, positing species transmutation across the whole spectrum of living things because he could see "no possible means of drawing [a] line, & saying here you must stop." This was a hint about his view of human origins. Although human evolution from other animals wasn't explicitly asserted in the book itself, it was provocatively suggested. Lyell had already read the proofs and seemed staggered by the implications, Darwin said, but had been helpful with criticisms and supportive overall. Lyell was a brick. Darwin hoped Hooker would give candid feedback in the same vein.

By now Darwin was getting some rest and hydropathy at

another watery resort, Ilkley Wells, at the edge of a moor up in Yorkshire. The place offered billiard tables, like Moor Park, and a few notably good players who dazzled him with their skills at "the American game." Some of them, he enthused to his son William, could make breaks of thirty or forty balls. The American game would have been one of several variants ancestral to what we call pool. And so, at this point, I suggest an imaginative pause to appreciate the tableau: Charles Darwin, having just completed the most consequential work of his life, seeking respite in the northern boondocks with a cue in his hand. He was known to smoke an occasional cigarette, for relaxation, in lieu of snuff, and maybe Ilkley Wells allowed that vice in the billiards room. Darwin takes a slow drag, holds it pensively, exhales. Squints through the smoke. Lays his smoldering cigarette fastidiously into an ashtray (certainly not on the edge of the table) and leans down. Crooks his index finger, sets his bridge. Six ball, gentlemen, he says; in the corner pocket. Tap . . . click . . . plop. "You cannot think how refreshing it is to idle away [a] whole day," he told Hooker, "& hardly ever think in the least about my confounded Book, which half killed me."

That confounded book was now called *On the Origin of Species by Means of Natural Selection, or the Preservation of Favoured Races in the Struggle for Life.* The title wasn't elegant but at least John Murray had convinced him to drop the word "Abstract." Optimistically, Murray ordered a first printing of 1,250 copies. The printer's bill for all those revisions on the proofs came to £72, a sizable chunk of cash, but Murray waived his right to deduct that from Darwin's royalties, correctly foreseeing that cordial relations with this author would be worth more in the long run. Although the writing process had been torturous to Darwin—an almost hysterical fit of work after so many years' delay—the blessings of forgetfulness came quickly and like a balm. He was still in Yorkshire when a first copy of the book reached him.

Holding it in his hands, he felt an irresistible flush of satisfaction, his private reward between the anguish of creation and the anguish that would come with the book's reception. He wrote immediately to Murray: "My dear Sir, I have received your kind note & the copy: I am *infinitely* pleased & proud at the appearance of my child." Abominable or not, it was his offspring, and the pride of parenthood covers a multitude of misgivings. But Darwin had grounds for his satisfaction. Without yet knowing it for certain, without claiming it, without having enjoyed it, he had dashed off a magnificently potent book that would change the world.

On November 22, several days before its official publication date, Murray offered it to booksellers. Based on what little they had heard about the contents, and on Darwin's reputation, they snatched the book up, ordering 1,500 copies against the 1,111 copies (first printing minus promotional giveaways) that were available. This is the precise reality behind a loose statement sometimes made—that the first edition sold out on the first day. It more than sold out, yes, at the wholesale level. Trickling into the hands of individual readers took longer. Still, the trade sale was a very strong start. Emma sent the news in a note to William, at Cambridge, adding: "Your father says he shall never think small beer of himself again & that candidly he does think it very well written." Murray promptly asked Darwin to get busy on a new edition, so they could reprint with some value added. Darwin got busy, making small corrections and revisions on the one copy he had in his hands.

November 22, 1859, was a Tuesday. It's an interesting day to consider, representing a nexus point between all the private turmoil that went into creating Darwin's book and all the public turmoil that has followed from it. As of that date, based on Murray's advance orders, *On the Origin of Species* was a commercial success, though few people had read it. Among those who had, reactions were mixed. A pre-publication review in one prominent journal, the *Athenaeum*, was caustically negative in a way that probably helped stimulate interest: "If a monkey has become a man—what may not a man become?" Never mind that nowhere in this work did Darwin say that monkeys had changed into men. He barely alluded to the subject of human origins. Oversimplification and scandalmongering had begun even before the book hit the stores.

32

It's safe to say that *The Origin of Species* (as it became known after Darwin himself dropped the *On* from a later edition) is one of the most influential books ever written. What printed works surpass it in reach and impact? Maybe the Bible, the Qur'an, the *Mahabharata*, and a few other scriptural texts that have inspired millions of people to piety and bloodshed. What stands in the same category? Maybe *De revolutionibus orbium coelestium* by Copernicus, maybe Newton's *Principia*, and maybe, if you count journal articles, the two papers of 1905 and 1916 in which Einstein described special and general relativity. Unlike those other great works of science, though, *The*

Origin is a book written in plain everyday language and meant by its author to speak to any attentive reader. Some of its grammatical constructions are a bit sinuous in the Victorian style; much of its writing is clear and crisp. Darwin was inconsistent as a literary stylist, sometimes bad, sometimes good, but even when bad he wasn't esoteric. Occasionally he just tried to put too much into a single sentence, a run-on construction with syllogistic premises, qualifications, facts, stipulations, and conclusions all linked together by semicolons and dashes like a giant protein molecule folding back on itself. Once in a while he wrote something beautiful and brilliant. Mostly he was an amiable explainer and narrator presenting one of the more astonishing tales ever told.

Although *The Origin* is the founding text of evolutionary biology, you can get a Ph.D. in that field at many American universities, and probably at British ones too, without having read it. Such neglect of a seminal document is oddly shortsighted, given that evolutionary biology itself is a historical science, concerned with examining and understanding the past as well as the present. The study of evolution proceeds from observed facts and found data more than from controlled experimentation. It still relies on Darwin's ideas and Darwin's terminology —most notably, on the idea and term "natural selection"—but its professional courses of training generally don't require students to read Darwin. That's too bad, because reading Darwin can be fun, even thrilling, as well as instructive.

It isn't always. In the course of his working life as a naturalist (an "amateur" naturalist, in the sense that he never held a job of any sort, let alone a scientific appointment) and a freelance writer (who liked making money from his books, though he didn't need to), Darwin wrote his share of somniferous duds. The harder and longer he labored, it seems, the more likely he was to produce a big, boring tome, chockful of carefully gathered facts, judiciously framed questions, and arcane conclusions, all presented with relentless lack of economy or flow. *The Variation of Animals and Plants under Domestication*, published in 1868, is no page-turner. Nor can I much recommend *The Different Forms of Flowers on Plants of the Same Species*, which came out in 1877. Some of his shorter books (they're not *very* short), such as *Insectivorous Plants* and *The Various Contrivances by Which Orchids Are Fertilised by Insects*, contain nice samples of one of Darwin's most appealing literary modes—his close, gentle examination of quirky creatures embodying large biological themes. But are those books urgent and compelling? Are they lively, readable works overall? No. His very last book, *The Formation of Vegetable Mould, through the Action of Worms, with Observations on Their Habits*, is a pleasant little surprise, largely because it's so unpretentious and eccentric. His barnacle volumes can't be counted against him, since those were always intended for the reference of specialists. The *Journal of Researches* (or, under its modern title, *The Voyage of the Beagle*) is the most accessible of his books, a colorful outpouring of narrative and description in the voice of a curious, unassuming young man; but it doesn't carry his strengths as a mature, conceptualizing scientist. His autobiography was written as a private memoir, for the family, and never published during his lifetime. *The Descent of Man* is really two books smooshed into one, as its full title admits: *The Descent of Man, and Selection in Relation to Sex*. The smooshing is far from seamless; there's a big, lumpy transition after the first seven chapters, right where *Descent* gives way to *Sex*. Humanity's descent from an animal lineage was one of Darwin's boldest ideas, true, but that book on the subject isn't one of his best efforts. Published in 1871, intended as a complement to *The Origin of Species*, it doesn't have the same sharp focus, inexorable momentum, and magisterial power. It sold well at the time, thanks to the notoriety of his ideas, but it doesn't reward attention nowadays as much as *The Origin* does.

Haste and anxiety seem to have been good for Darwin, at least as a writer, at least in the one crucial case. Alfred Wallace, by shocking him with the threat of preemption, forcing his hand, inadvertently did him a big favor. The quick-and-dirty "abstract" proved to be readable, popular, persuasive, and efficacious in a way that the big book on natural selection wouldn't have been. That big book remained unfinished by Darwin, and unpublished during his lifetime, partly because he had lost interest in its grand schema of encyclopedic exposition and partly because *The Origin* had rendered it unnecessary. He did salvage the first two chapters, converting them into his book on domestic variation. The rest of the long manuscript, comprising eight and a half chapters, didn't see daylight until 1973, when a scholar named R. C. Stauffer edited it for publication under the title *Charles Darwin's Natural Selection*. Although Stauffer's edition is useful as textual background, its main value lies in making *The Origin* look good by contrast. It shows how lucky Darwin was—and readers too—that Alfred Wallace barged in when he did.

The littler book, Darwin's abominated volume, has a more complicated editorial history. Let's say you've decided to read or to reread *The Origin of Species*. Immediately you face another decision: Which *Origin*? Six different editions appeared in England during his lifetime, each of the latter five showing revisions by the author. Many of those revisions were substantive. Sentences were added and others cut, passages were rewritten for clarity, second thoughts replaced first thoughts and then third thoughts replaced those, qualifications were inserted, fudges were factored, whole blocks of new argument answered criticisms of the more astute reviewers. Another literary scholar, Morse Peckham, assembled all the changes made throughout sequential editions into a "variorum text" of *The Origin*, published in 1959 on the book's centenary. From Peckham we know, for instance, that for the hurried second edition demanded by John Murray after the sellout, 9 sentences were dropped, 30 were added, and 483 received some tinkering. To the third edition, in 1861, Darwin attached what he called "An Historical Sketch of the Recent Progress of Opinion on the Origin of Species"—that is, a survey of his intellectual predecessors—in response to accusations that he was taking credit for ideas that others had published before him. This sketch, placed as a preface, acknowledged earlier thinking by Lamarck, Geoffroy, his old Edinburgh mentor Robert Grant, the still anonymous author of *Vestiges*, the anatomist Richard Owen, an obscure writer on shipbuilding timber named Patrick Matthew (who'd gotten jealously miffed over Darwin's natural selection, insisting he'd mentioned the same idea back in 1831), and his own grandfather Erasmus, among others.

For the fourth edition, in 1866, Darwin added two pages of further credits to the historical sketch and expanded the main text by 10 percent, including an enlarged section on embryology and development. The fifth edition was the first in which Darwin adopted Herbert Spencer's famous phrase, "the Survival of the Fittest," as a rough synonym for natural selection.

All these later editions incorporated more supporting facts and examples, some drawn from the recent work of scientists inspired by *The Origin* itself as the cycle of evolutionary theory and research began to turn round. Darwin's text was evolving, adapting, in reaction to the ongoing debate it had triggered. For the sixth edition he included a whole new chapter, "Miscellaneous Objections," answering one of his most aggressive critics point by point. This was the edition, too, for which Darwin struck the word "On" from his title, so that *On the Origin of Species by Means of Natural Selection* could henceforth be accurately shortened, in casual speech, to *The Origin of Species*.

The sixth edition, published by Murray in 1872, was the last worked on by Darwin himself, and for that reason has often been taken as the authoritative text, the one people should read if they want to know what Darwin *really, finally* meant to say. In my non-expert opinion, this is misguided. The last version wasn't necessarily the best, the most interesting, or the most significant. It wasn't even, to revert to that problematic word, the most "Darwinian." Not all the changes Darwin made to *The Origin* between 1859 and 1872 were improvements, and not all his improvements were as consequential as what he'd said, most daringly, in the book as first published. After thirteen years of extraordinary international hubbub, its main ideas had become widely known, from secondhand reportage and literary chatter as well as from the text itself. Those ideas were now swirling throughout the scientific communities of Britain, continental Europe, and America, and swirling also (more blurrily) in the public awareness, independent of whatever subtle revisions Mr. Darwin made for his latest edition. Evolution by natural selection was, by 1872, a bigger intellectual phenomenon than *The Origin of Species*, whereas in November 1859 the book was (with all due respect to Wallace) the definitive embodiment of the theory. Fiddle as he might with refinements, rebuttals, and adjustments of emphasis, Darwin wasn't perfecting or undoing what he had initially done, which was to startle the world with a new way of seeing life on Earth.

So my advice is, ignore the afterthoughts. Ignore the paperback reprints of the sixth edition. Trust no one; before you buy, before you read, check the small print in front under "Note on the Text," or the discreet line of dating opposite the title page. Whether you turn to *The Origin of Species* as a scientific work or as a historical and literary document, do yourself and Charles Darwin a favor: Find a reprint (preferably a facsimile reprint, with the original typeface and pagination) of the first edition. That's the book, with all its courageous freshness and its flaws, that provoked the most cataclysmic change in human thinking within the past four hundred years.

33

The book opens in a mild, unassuming tone of reminiscence:

> When on board H.M.S. 'Beagle,' as naturalist, I was much struck with certain facts in the distribution of the inhabitants of South America, and in the geological relations of the present to the past inhabitants of that continent. These facts seemed to me to throw some light on the origin of species—that mystery of mysteries, as it has been called by one of our greatest philosophers.

The passage reads easily but there's much alluded to: biogeography, paleontology, closely allied species adjacent in time. For the "mystery of mysteries" phrase from John Herschel, Darwin had dipped back into his transmutation notebook "E," using an entry he'd made as an excited young man in December 1838. The part about throwing "some light" is a whopping understatement, which he will repeat to good effect, even more whoppingly understated, at the end of the book. Beyond touching those points, his other smart choice here is to come sailing into view—offering himself to our first impression—aboard the *Beagle*. It makes him seem personable and grounded in experience; back in 1859, it also reminded readers that this middle-aged theorist was the same fellow who, twenty years earlier, had delivered the popular travel narrative, *Journal of Researches*.

Darwin's six-page introduction, begun with that sentence, is a sort of overture in which he sounds his major themes: that "innumerable species inhabiting this world have been modified," giving them amazing "perfection of structure and coadaptation," and that the mechanism he calls "natural selection" can account for those modifications. He doesn't use the word "evolution." (That term appears nowhere in the first edition, though Darwin does conclude the book's very last sentence by saying that many wondrous species "have been, and are being, evolved.") The old, familiar, and provocative word "transmutation" is omitted here, too. Instead he talks in these opening pages about "modification and coadaptation" (and, later in the book, about "descent with modification" or his "theory of descent"). The other important theme announced in his introduction is the "struggle for existence," the same phrase for the same idea that Alfred Wallace hit upon independently. And of course he mentions Malthus.

Then he writes:

> As many more individuals of each species are born than can possibly survive; and as, consequently, there is a frequently recurring struggle for existence, it follows that any being, if it vary however slightly in any manner profitable to itself, under the complex and sometimes varying conditions of life, will have a better chance of surviving, and thus be *naturally selected*. From the strong principle of inheritance, any selected variety will tend to propagate its new and modified form.

Add just two further ideas—that continued selection leads to prodigious extremes of adaptation, and also eventually to divergence of lineages—and this could stand as the abstract of the abstract.

The main text is divided into fourteen chapters. The peculiar sequence of those chapters reflects a counterintuitive decision by Darwin: to discuss the mechanism by which evolution occurs before calling attention to the consequent phenomena. That is, he works first to persuade readers about natural selection—that it *can* happen and *must* happen, given variation and the struggle for existence. Only afterward does he offer evidence showing that evolution itself, by whatever mechanism, *has* happened. This may seem backward to us, but in 1859 it was shrewd, given that species transmutation was a familiar idea that had been proposed and rejected previously, whereas natural selection was the breakthrough concept that could make transmutation plausible, even irresistible, to skeptics.

Darwin devotes his first chapter to variation among domestic animals and plants, noting how much of it routinely occurs, and how breeders use such small differences to transform lineages of livestock, pets, and crops. Dogs, cows, pigs, goats, strawberries, potatoes, dahlias, hyacinths, rabbits, sheep, horses, ducks—every domesticated species encompasses variation. One chicken, if you know your birds, is not quite like another. A good pig differs from a mediocre pig. And so also, of course, with pigeons—especially fancy pigeons, his favorite case. Drawing on the expertise gleaned from his own coops, his esoteric reading, and the London fanciers' clubs he occasionally visited, Darwin argues that all pigeon breeds—the tumbler, the fantail, the pouter, and many others—are descended from one species of wild dove, *Columba livia*. What explains their extravagant forms of differentiation? The answer is selection, as practiced by humans. What explains racehorses versus dray horses, greyhounds versus bloodhounds? Again, selection by breeders. Nature somehow provides the small variations. Humans select preferred variations when they pair their animals or pollinate their plants. Those preferred variations are perpetuated—and magnified, through accumulation—over generations of domestic breeding. The result is specialized forms, differing from their ancestors in ways that people find useful or amusing. *Artificial* selection: the first leg of Darwin's cardinal analogy.

Then he turns to variation in the wild. "No one supposes that all the individuals of the same species are cast in the very same mould," he says. Anyone who looks closely will admit that wild animals within any species, like domestics, differ slightly from one another. Wallace saw it among his boxes of beetles, birds, and butterflies collected for sale. Darwin saw it among his infernal barnacles. Variation is exactly what made taxonomy so damned difficult. But most people in 1859 assumed that such differences, among wild creatures, are limited and evanescent. Other naturalists besides Darwin and Wallace had noticed them, but considered them unimportant. If species are immutable, according to this view, then variations are minor wobbles around the archetypal essence of each species, to which every digression will eventually return. Varieties within a species are populations of such wobblers, insignificant and impermanent anomalies constrained by uncrossable boundaries of species identity.

No, it isn't so, says Darwin. Varieties can't be so easily dismissed. In fact, even defining the two words—"species" and "variety"—is a tricky task. The tricky part is distinguishing one of those categories from the other, and the hardest thing about classifying specimens within such a category is that sometimes ambiguity can't be resolved. Lines blur. Taxonomists disagree. One botanist will look at a large group of plants and see 251 species; another expert will look at the same group and see only 112 species, plus 139 false or trivial distinctions. From his Galápagos experience and the trouble afterward in classifying the birds, Darwin remembers being "much struck how entirely vague and arbitrary is the distinction between species and varieties." The real distinction, he concludes, is this: a difference in degree. Species within a genus differ more than—but in the same ways as—varieties within a species do. The minor differences between varieties can accumulate, until they become the major differences between one species and another. It's the very point that headlined Wallace's 1858 paper, "On the Tendency of Varieties to Depart Indefinitely from the Original Type." Darwin didn't need to see that paper (and he certainly didn't want to) because he had arrived at the same insight himself.

These opening chapters represent Darwin's direct assault on the old mode of thinking, which held species to be divinely created and fixed, like ideas stored in the file cabinet of God. The two chapters lay a groundwork for his discussion of natural selection, but they do more. They deliver one of his most profound contributions to scientific thought. As the population geneticist Richard Lewontin has lately written: "Darwin revolutionized our study of nature by taking the actual variation among actual things as central to the reality, not as an annoying and irrelevant disturbance to be wished away." He allowed us to see the living world as endlessly various. He helped us understand the whole physical universe as a realm of concrete contingencies, not imperfectly represented ideals.

Chapter III is his treatment of the struggle for existence, using Malthusian arithmetic and empirical data to dispel the notion of nature in a state of divinely ordered tranquility. Nature's real order *isn't* peaceable; it's a desperate scramble, even when the desperation is muted and obscure. Darwin alludes to a famous statement by the Swiss botanist A.-P. de Candolle, suggesting that "all nature is at war, one organism with another, or with external nature." Predation, competition, parasitism, overcrowding. As species continue to procreate, there just isn't enough food or enough space for their offspring. Reproductive rates are geometric. Habitat is finite. Fortunately, many dangers loom. If struggle and death didn't sweep away most individuals of most species, the lands, seas, and skies would be impossibly full. Humans are relatively slow breeders, but the arithmetic holds true even for us: If we all bred and survived, doubling our population every generation, within a few thousand years there wouldn't be standing room for another person on the planet. Elephants breed even more slowly. But their inherent rate of increase is geometric, too. A single female elephant, if she produced just six offspring in her lifetime, and if every descendant reproduced similarly, would in five centuries (according to Darwin's rough calculation) yield a population of 15 million. That's a lot of elephants. Too many. It doesn't happen. Why not? Because every elephant must struggle—to survive, to reproduce—and many fail.

Taking an image from another of the old notebooks, Darwin writes: "The face of Nature may be compared to a yielding surface, with ten thousand sharp wedges packed close together and driven inwards by incessant blows, sometimes one wedge being struck, and then another with greater force." This is the struggle for existence. Something's gotta give. Just as not every wedge can fit, not every creature can find its place, meet its needs, achieving survival and reproductive success. Now consider, Darwin says at the start of *The Origin*'s chapter IV, how this struggle interacts with the fact of variation.

Chapter IV is the book's core, in which he makes explicit his great analogy between domesticated varieties and wild species. If selective breeding by humans can create such peculiar modifications, Darwin asks, "what may not nature effect" through natural selection?

Imagine, say, an island filled with native creatures. Imagine that it undergoes a change of climate. The new climate presents new difficulties to the natives. Making matters worse for them, immigrant creatures begin invading across the water. "In such case, every slight modification, which in the course of ages chanced to arise, and which in any way favoured the individuals of any of the species, by better adapting them to their altered conditions, would tend to be preserved; and natural selection would thus have free scope for the work of improvement." The "improvement" could take strange forms. Depending on circumstances, it might yield giant tortoises, pygmy deer, huge flightless birds, arboreal kangaroos, iguanas that dive for seaweed, mammoth cockroaches, or seed-cracking finches with the oral equipment of grosbeaks.

Furthermore, Darwin argues, natural selection leads to more than just small changes and nifty adaptations. It leads also to widening divergence between groups of creatures—between varieties, between species, between genera and higher categories—and thereby to the enormous diversity of life on Earth. That diversification is what allows such an abundance of individual creatures, and of different *kinds* of creatures, to co-exist within a small patch of forest, on an island, or in a little pond. As an instance, he mentions a single patch of turf, just three feet by four feet, that he examined himself. This patch had stood exposed and undisturbed for many years. Taking a complete census of the plants, Darwin found twenty species, representing eighteen different genera within eight orders. How could they all survive on such a small rectangle of dirt? They survived because they differed sufficiently from one another—in the ways they sought light, water, mineral nutrients, and reproductive success—and those differences minimized competition. Divergence is the phenomenon that allows greater amounts of biological success to be extracted from a finite amount of physical resources.

Given the Malthusian scramble for those resources, Darwin says,

> The more diversified the descendants from any one species become in structure, constitution, and habits, by so much will they be better enabled to seize on many and widely diversified places in the polity of nature, and so be enabled to increase in numbers.

With that sentence, especially its phrase about "places in the polity of nature," written before the word "ecology" even existed, he prefigures the concept of ecological niche.

The book's middle chapters are devoted to complicated topics such as instinctual behavior, hybrid sterility (which helps preserve differences between populations once they arise), and Darwin's solution to the apparent problem of transitional structures. The last of those is an old objection against his theory, still sometimes raised by creationists. What he means by a transitional structure is one not fully developed to its higher potential—for instance, a slightly winglike appendage that's useless for flying, or the primitive precursor of an eye. The problem lies in understanding how natural selection could produce such half-baked pastries. If variations occur in tiny increments, and natural selection preserves only the advantageous ones, then what conceivable advantage adheres to each incremental change during the transitional phase—when a proto-wing is not yet aerodynamic, and a proto-eye cannot yet focus an image?

Darwin answers with examples of transitional forms that *are* adaptive (for instance, the "wings" on a flying squirrel or a flying fish) and with careful logic about the value of light-sensitive organs such as the ocelli in some invertebrates, which aren't fully developed eyes. He also suggests that the *type* of advantage may change fortuitously, from one opportunistic use to another, as a structure evolves. As illustration, he describes an obscure pair of structures known as the *ovigerous frena*, found among some barnacle species (the stalked ones) as two tiny folds of skin. These frena, these folds, secrete a sticky goo that helps hold gestating eggs within the body sack. The same basic organs appear in transmuted form among other barnacle species (the sessile ones), for whom they serve a different adaptive purpose, related to breathing. Lo, the two sticky folds have become branchiae. Darwin presents this case with sublimely erudite confidence, since he discovered and named the ovigerous frena himself. Who *says* the barnacle years were wasted?

34

By this time you're deep into the book. Only there, beginning with chapter IX, does Darwin shift his focus from the mechanism, natural selection, to the phenomenon, evolution. His tactical approach changes, too. Instead of arguing that natural selection *must* occur, he offers evidence that evolution *has* occurred. His evidence falls mainly within four categories: biogeography, paleontology, embryology, and morphology.

Biogeography, as you've seen, was the starting point of his own conversion to evolutionary thinking, and the guiding inspiration for Wallace, too. It's a vivid field of study, big and gaudy as all outdoors, but amid the sheer pageantry there's much buried meaning. Anyone who considers the geographical distribution of animals and plants, Darwin writes, must be struck by the clustering patterns among similar forms. Several species of zebra in Africa, none elsewhere. Several species of kangaroo in Australia and New Guinea, none elsewhere. Old World monkeys (the catarrhines) only east of the Atlantic Ocean, New World monkeys (the platyrrhines) only west of it. Many lemur species in Madagascar and its nearby islands; no lemurs anywhere else. Many toucan species in Central and South America; no toucans anywhere else. Where lemurs and toucans are absent, although habitat and climate might suit them, their roles are filled by other sorts of species—monkeys instead of lemurs in Africa, and hornbills there instead of toucans. Why? Are these patterns just accidental, or do they tell a story?

Darwin quotes Wallace's 1855 paper, the one whose meaning he missed on first reading, to the effect that "every species has come into existence coincident both in space and time with a pre-existing closely allied species." Mr. Wallace and I now agree, he says, that this coincidence is explained by descent with modification, every species diverging from another in space and time. Adjacent areas of South America are inhabited by two similar species of large, flightless bird (the rheas), not by ostriches as in Africa or emus as in Australia. South America also has agoutis and bizcachas (largish rodents) in terrestrial habitats plus coypus and capybaras in the wetlands, rather than hares and rabbits in terrestrial habitats plus beavers and muskrats in the wetlands, like North America. Why not the same critters everywhere? Why should closely allied species inhabit neighboring areas on each continent? And why should similar habitat on different continents be occupied by species that aren't so closely allied? "We see in these facts some deep organic bond, prevailing throughout space and time," Darwin says. "This bond, on my theory, is simply inheritance." Similar species occur nearby in space because they have descended from common ancestors.

Paleontology reveals other clustering patterns, in the dimension of time. Look at the fossilized mammal bones and the living mammals of Australia, Darwin suggests. Look at the ancient and less ancient giant birds of New Zealand. Look at the fossil snails from Madeira and the species living there now. Are these similarities, between older and newer species, incomprehensible? Are they random happenstance? No. "On the theory of descent with modification," he says, the occurrence of similar but not identical species during different geological eras within the same area "is at once explained."

Embryology too involves patterns that beg for explanation. Why does the embryo of a mammal, passing through its developmental phases, spend one phase resembling the embryo of a reptile? Why does it, at another point, show gill slits like the embryo of a fish? And since embryology in a broad sense considers immature growth stages, not just unborn or unhatched forms, those questions lead to others. Why do lion whelps have striped legs, like the grown-ups of their close relative, the tiger? Why is a larval barnacle, swimming freely before metamorphosis, so similar to the larva of a brine shrimp? Why do the larvae of moths, flies, and beetles resemble one another (they're all wormy) more than any of them resembles its respective adult? Because, Darwin writes, the embryo is "the animal in its less modified state," and that state "reveals the structure of its progenitor."

Morphology, the study of anatomical shape and design, is in Darwin's words the "very soul" of natural history. What could be more suggestive than that the hand of a human (shaped for grasping), the paw of a mole (shaped for digging), the leg of a horse (shaped for running), the fin of a porpoise (swimming), and the wing of a bat (flying) should all reflect an underlying five digit pattern, with modified versions of the same bones in the same relative positions? Darwin doesn't claim to be the first naturalist to notice such homology; he reminds us that it was crucial to Geoffroy's formalist vision, which saw a "unity of plan" beneath the multiplicity of animal shapes. And the "same great law" of homologous parts can be recognized also in the mouths of insects. The long spiral proboscis of a moth, the folded snout of a bee, the fierce jaws of a beetle—they're made for different purposes but from common elements: mandibles, maxillae, an upper lip. The different parts of a flower, such as stamens and pistils, sepals and petals, are likewise homologous from species to species. What's the reason behind this recurrence of a few basic designs? Under the old view of special creations, Darwin notes, the only answer was "that it has so pleased the Creator to construct each animal and plant" with stingy economy of invention. That answer didn't really make sense. Why should an all-powerful deity economize? Darwin's answer is descent with modification. Homologies reflect the fact that natural selection—which isn't omnipotent and *is* economical, limited by history and circumstance—works upon patterns passed down from ancestral forms.

One use of morphology—the use to which Geoffroy and Cuvier put it—is systematic classification, the task of sorting all species into groups within other groups. Darwin, the former barnacle taxonomist, considers this topic worth twenty-three pages. The grouping of species into larger categories isn't arbitrary, he says, like sketching stars into constellations for human amusement or ease of memory; biological groupings are assumed to have some deeper basis. But what? Classifiers try to arrange species, genera, families, and other sets into a system that's not just handy for reference but somehow "natural" and objective. You can see those arrangements applied in the layout of any zoo. Here are the monkeys, there are the big cats, and in that building are the alligators and crocodiles. Birds in the aviary, fish in the aquarium. But what about porpoises and manatees? Clearly they're mammals, not fish, yet fishlike in their habitat needs—so do they go into the aquarium, too? Zoo designers may blur lines for the sake of practical convenience; taxonomists, on the other hand, do their best to recognize fundamental resemblances rather than superficial ones. For instance, all vertebrate animals (by definition of the category) have backbones. Among vertebrates, the mammals have fur and mammary glands, not feathers or scales. Among mammals, the marsupials have pouches in which they carry and nurse their dependent young. Among marsupials, the kangaroos have big feet and strong tails. What's the ultimate source of this orderliness? Many naturalists of his day, says Darwin, believe that a good system of classification simply "reveals the plan of the Creator." That explanation doesn't satisfy him. It might or might not be true but it offers nothing in scientific knowledge. Darwin's alternative is that "all true classification is genealogical." Commonality of descent, he says, is "the hidden bond which naturalists have been unconsciously seeking." Alligators resemble crocodiles because they derive from common ancestors, not because of some divine choice to create multiple species of big aquatic reptiles with cone-shaped teeth.

Rudimentary organs are still another form of morphological evidence, illuminating to contemplate because they show that the living world is full of small imperfections: the blind eyes of a cave fish, the wing stubs of a kiwi, the appendix of a human. In a sense these are transitional structures too. But Darwin considers them separately from flying-fish wings, insect ocelli, and ovigerous frena in barnacles because rudimentary organs (he also calls them "atrophied" or "aborted") seem to mark stages in a course of evolutionary deterioration (localized deterioration, that is, reducing the organ but not harming the creature overall) rather than stages in evolutionary improvement. If that isn't the explanation for their presence, and their odd uselessness, what is? For his readers Darwin raises the same question here, late in *The Origin*, that intrigued him privately back in notebook "B": Why do men have nipples? And why do some snakes carry the rudiments of a pelvis and hind legs buried inside their sleek profiles? Why do certain species of flightless beetle have wings, sealed within wing covers that never open? Such vestigial features stand as remnants of the history of a lineage.

Just what makes rudimentary organs shrink away? That's a complicated issue—more complicated than even Darwin knew. He suspected that *disuse* of such organs is sufficient cause; but modern evolutionary theory (and I'll come back to this) says he was wrong. Okay, nobody's perfect. Charles Darwin certainly wasn't. He had an appendix, he had nipples, none of which served any useful purpose, and he occasionally made mistakes, even in *The Origin of Species*. Anyway, whatever impels the deterioration of elaborate organs into rudiments, the result is a record of evolutionary change.

In his final chapter, Darwin declares that the entire book is essentially "one long argument" linking the idea of common descent to the idea of natural selection. After recapitulating his main facts and inferences, he rises to a rhetorical crescendo, forecasting accurately that his theory will ignite "a considerable revolution in natural history." The job of systematists, given this new perspective, will become easier and less ambiguous. Fresh inquiries will be opened into the causes and laws of variation. Natural history will be altogether more interesting. The study of domesticated species will gain value. Paleontology will be clarified, biogeography will advance, embryology and the scrutiny of rudimentary organs will reveal connections between living species and ancient prototypes. "In the distant future," Darwin writes, "I see open fields for far more important researches." Psychology, for example—the origin of mental powers will be understood in an entirely new way. And then he lets drop his most famously coy remark: "Light will be thrown on the origin of man and his history." Still being cautious, even in his revolutionary manifesto, Darwin says nothing more—for now—about human evolution.

Instead he shifts to another touchy subject: the ways of God. Many eminent authors, he admits, are satisfied to believe that each species has been specially created. Not Darwin. "To my mind it accords better with what we know of the laws impressed on matter by the Creator," he declares, "that the production and extinction of the past and present inhabitants of the world should have been due to secondary causes, like those determining the birth and death of the individual." There's a big theme, and a deep conviction, buried in that phrase about "*laws impressed on matter*"—bigger and deeper even than the subject of evolution. Darwin believed that the universe was governed by fixed laws, not by capricious divine interventions. He was still enough of a theist, at least in 1859, to write of "the Creator" as ultimate source of those laws, but his whole intellectual life was grounded in the confidence that such laws were discoverable and unchanging. He hinted as much, back at the start of *The Origin*, with a small quotation from William Whewell placed as an epigram opposite his title page:

> But with regard to the material world, we can at least go so far as this—we can perceive that events are brought about not by insulated interpositions of Divine power, exerted in each particular case, but by the establishment of general laws.

Now, in the book's final paragraph, Darwin returns to that big theme. Think of evolution as the result of fixed laws, he urges: like gravity, or the movement of heat. Evolution's governing laws include biological growth, reproduction, inheritance, variation, population pressure, and the struggle for existence, all combining to yield natural selection, divergence, and the extinction of less adapted forms. From the war of nature comes an exalted result: the higher animals. Isn't that a more satisfying and majestic notion than requiring God personally to design every tick, clam, and flatworm?

To Darwin it is. "There is grandeur," he says finally, echoing phrases he first drafted for his 1844 essay, "in this view of life. . . ." The full, closing passage of *The Origin* is famous, but worth quoting again:

There is grandeur in this view of life, with its several powers, having been originally breathed into a few forms or
into one; and that, whilst this planet has gone cycling on
according to the fixed law of gravity, from so simple a
beginning endless forms most beautiful and most wonderful have been, and are being, evolved.

There *is* grandeur in that view. It's an eloquent conclusion
to a magnificent, hastily composed, compelling, and seriously
flawed book.

35

Closely rereading *The Origin of Species*, with less attention to
the core of its argument and more attention to the author's
voice, to his style of logic, to his omissions and mistakes, to
the scope of his claims, helps put his accomplishment into
perspective. A bit of modestly critical scrutiny reveals, without
disrespect to what Darwin achieved, that this great book is not
great on every page and in every way.

One of its defects is Darwin's incessant apologizing for the
fact that *The Origin* isn't three times longer. "This Abstract,
which I now publish, must necessarily be imperfect," he writes
in the introduction. "I cannot here give references and authorities for my several statements." We've heard that before, in the
anxious letters to his friends: Woe and alas, my stinking book,
it's just a miserable abstract, squashed and inadequate. But
now he's making a public excuse, not just fretting privately,
and Darwin frames the excuse in a disingenuous form. "I
much regret that *want of space*," he says on page 2, "prevents
my having the satisfaction of acknowledging the generous

assistance . . ." et cetera. The italics are mine. "If I had *space*," he claims later (again, my italics), "I could quote numerous passages to this effect" (never mind, about what) "from highly competent authorities." Later still, making a different point: ". . . but I have not space here to enter on this subject." So it's "want of space" that's the problem?

Actually, no. Darwin was free to take as much space as he needed. Page limits had been a genuine concern earlier, true enough—during the first months of composition, when he imagined that his abstract might be publishable as a journal article. Then he overflowed those limits and, making a tactical choice, decided that it must be a book. Midway through the writing, in late 1858, he projected that his volume would fill 400 pages, a rough guess later revised to 500. John Murray, his publisher, never specified any limit. But in the 1859 text as we find it published, he can't stop complaining about his self-imposed constraints. "I could show by a long catalogue of facts . . ."—yet he doesn't. "I shall reserve for my future work the discussion of these difficulties . . ."—but that work, the big book, never came. About comb-building behaviors among honey bees, he writes, "if I had space, I could show that they are conformable with my theory." Throughout the book he repeats that lament: *Can't give details, sorry. Maybe later.* The pretext, when he offers one, is: *No space.* What was Darwin really fussing about?

Not a shortage of space but a shortage of time. Alfred Wallace had scared the bejesus out of him, he knew he'd delayed too many years, and now he felt desperately rushed to get his book into print. Dignity prevented him from admitting that.

Another quirk in *The Origin* is the extent to which his "one long argument" relies on probability and personal attestation.

Taken properly within its philosophical context—the context of inductive science, as it had lately been outlined by Whewell, among others—this should probably be seen as a strength of the book, not a weakness. Darwin doesn't claim to *prove* the reality of evolution by natural selection. In fact, "proof" is a word that he seldom uses—and when he does, it's often negatively, to acknowledge some intractable ambiguity. About the notion that embryology gives glimpses of evolutionary lineage: "This view may be true, and yet it may never be capable of full proof." About the idea, sometimes asserted, that variation in the wild has strict limits: "the assertion is quite incapable of proof." More important, Darwin understands that good inductive science (which had become the ideal by the time he was writing) can never, unlike mathematics, prove a result beyond any logical possibility of doubt. Instead of claiming to *prove* his big theory, he moves the reader toward it persuasively by way of accreting evidence. The goal is to show that his hypothesis explains a larger and more interconnected collection of data, with greater probability, than any alternative hypothesis. Along the way, Darwin makes statements such as "I think it highly probable that" and "I am convinced that," buttressing the evidence with his own amiable persona as a fair-minded English gentleman to suggest that these conclusions can probably be taken as right.

This is a point with some relevance, in our own time, to the conflict between evolutionism and creationism. It's an arid truth, but one that the defenders of evolutionary theory (and the teaching of it in public schools) against religion-based political challenges would do well to remember. The complexities of epistemology, as well as those of biology, shouldn't get lost in the arguing. No, you can't *prove* that all species have evolved from common ancestral lines, with natural selection as a major driving mechanism, and Charles Darwin himself didn't claim that you could. It's just very, very probable that this explanation of the living world is correct, based on the evidence Darwin mustered and all that's been added since. The alternative explanations are either less probable within the realm of physical cause and effect, or else they're scientifically meaningless (because untestable against negative data) expressions of religious belief.

Besides lacking any claim of absolute certainty, *The Origin* is marked by some other notable omissions. As I've mentioned, it lacks the word "evolution." (That term carried undesirable connotations, in 1859, related to a sort of mystical unfolding or unrolling of forms.) It lacks a good explanation for the source of those crucial variations upon which selection acts. It lacks an unambiguous statement about whether such variations are haphazard or somehow directionally evoked. (The adjective "random" appears nowhere in the book, and to say that variations result from "chance" is, as Darwin admits, misleading. He does imply, though, that they are undirected.) Despite its attention to the principle of divergence, it lacks clarity on the key matter of how speciation—as distinct from adaptation—occurs. (When two populations of a species diverge from each other, what factor accounts at a certain point for their irreversible separation into two species?) It lacks insight into the mechanics of inheritance—the vital matter of how selected variations are passed along. Finally, the book lacks any explicit assertion that we humans share an ancestor with apes.

One thing the book doesn't lack is the idea that acquired characteristics can be inherited. Although that idea is sometimes considered synonymous with Lamarckism, it actually predated Lamarck's work and remained more enticing than the Frenchman's other propositions. In Darwin's plain language, it sounds concrete and sensible: the "effects of use and disuse."

"I think there can be little doubt that use in our domestic animals strengthens and enlarges certain parts," he writes in *The Origin*, "and disuse diminishes them; and that such modifications are inherited." Furthermore, it's not just domestic animals that show this trait, he says; wild ones do, too. "I believe that the nearly wingless condition of several birds, which now inhabit or have lately inhabited several oceanic islands, tenanted by no beast of prey, has been caused by disuse." The dodo of Mauritius, the cassowaries of New Guinea, the emus of Australia, and of course the kiwi, all surrendered their wings to this principle, he thought. Use 'em or lose 'em. A giraffe, in his view, contrary to Lamarck's (as Darwin construed it), couldn't simply *will* its way to a longer neck; but by the habit of stretching for high food, it could add increments of length, and those increments (here's his mistake) could be inherited. Muscles of a blacksmith, likewise. By their efforts and habits, individual creatures earn bodily improvements . . . and they can pass those improvements along to their offspring, Darwin believed.

These confusions and omissions suggest some of the unfinished scientific business remaining for Darwin, and for his acolytes and successors, when *The Origin of Species* first appeared. Darwin himself understood that this book he'd dashed off wasn't perfect. Although he foresaw a revolution in natural history, he recognized that his "abstract" was just the opening salvo, not the declaration of final settlement terms. He knew that the work of what we now call evolutionary biology had only started, and he meant to stay engaged as it progressed. He was still struggling to comprehend variation. He wanted an explanation of heredity. He intended to address the hot topic of human origins.

Meanwhile the book made him famous—far more famous than he'd been as a conventional naturalist and writer—and profoundly controversial. It was translated (badly and irresponsibly, in some cases, by foreign thinkers with their own agendas), published abroad in authorized and unauthorized editions, widely reviewed, admired, denounced, released in a cheap edition by Murray for a bigger market, and talked about by many more people than actually read it. It sold roughly 25,000 copies, of the English editions alone, during Darwin's lifetime. "The real triumph of Darwin's book came after his death," according to Morse Peckham, editor of the variorum text. "The profits of the American pirates must have been enormous." Those numbers are unavailable, as are totals reflecting the book's global reach. A bibliographical checklist, published in 1977, recorded 425 distinct editions of *The Origin of Species* (not counting reprints of each edition) just to that point, including four in Hungarian, two in Hebrew, two in Romanian, two in Latvian, four in Korean, one in Hindi, and fifteen in Japanese. Darwin himself devoted a sizable share of his energies, over the dozen years following first publication, to revising it, promoting it (he mailed off quite a few complimentary copies), monitoring its reception (yes, he read his reviews), and playing his role (mainly by letter) in the scientific discussion it provoked. The book succeeded hugely in some ways, and failed in others. It made evolution seem plausible. But it left many of Darwin's scientific colleagues—never mind lay readers and religious critics—unwilling to accept natural selection as the mechanism. That idea was still too big, too scary, too cold.

For better and worse, with its flaws and its majesties, *The Origin of Species* stands as Darwin's ultimate statement of his theory, and the 1859 edition articulates that statement most freshly and boldly. It was never supplanted by the encyclopedic tome, *Natural Selection*, that he had intermittently intended to write. Through those five later editions of *The Origin*, he continued monkeying with his text, sometimes improving it but often just adding confusion, caution, and unnecessary words. By 1869, he sounded weary of that. "If I lived twenty more years and was able to work, how I should have to modify the *Origin*," he confided to his dear old friend Hooker—and then, changing moods, mustering resoluteness, "how much the views on all points will have to be modified!" But he wouldn't live twenty more years. Nor did he expect to. So he sighed: "Well, it is a beginning, and that is something. . . ."

It was something indeed: less than the book he had hoped to write, but more than enough to cause a ruckus.

The Fittest Idea

1860 to the future

36

Most people today don't realize that natural selection, Darwin's greatest and most troubling idea, fell into disfavor among evolutionary biologists for fifty or sixty years. Most people imagine that the Darwinian revolution, so called, was a relatively quick campaign fought and won in the late nineteenth century. It wasn't. It was an up-down-up scuffle for decades.

There were two principal questions, contested almost independently: (1) Has evolution occurred? and (2) Is natural selection its main causal mechanism? Despite some horrified outcries from religious leaders and pious scientists, the descent of all species (even humans) from common ancestors became widely accepted rather soon after *The Origin of Species* was published. Despite his careful arguments in the first half of the book, Darwin's hypothesis as to the causal mechanism did not. Why not? Because the idea of natural selection seemed profoundly materialistic and gloomy—that is, it was both literally and figuratively dispiriting—whereas the idea of evolution seemed merely insulting (as applied to the human species) and bizarre. Evolution contradicted William Paley's natural theology, as propounded in 1802, yes; but Paley's natural theology was an ingenuous, prescientific vision that had outlived its time (except in America, where it returned during the late twentieth century under the label "Intelligent Design") and was soon supplanted by the idea that species, rather than being individually created by God, had somehow evolved one from another. Natural selection struck deeper, undermining the whole notion of godly purpose. Evolution could be reconciled with belief that a divine Creator had established laws governing the universe, had set life into motion, had allowed species to change over time, and then— at some magical moment—had injected a unique spiritual dimension into the primate species that was later to be known (by its own self-naming) as *Homo sapiens*. Natural selection, on the other hand, seemed to preclude that belief. It did, anyway, if taken strictly and seriously—the way Charles Darwin took it.

The crux of the matter was not natural selection itself but the variations upon which it works. What causes those small differences between parents and offspring, and between one competing individual and another, which serve as the raw material for adaptive change? What laws govern their scope, rate of occurrence, and character? Are they purely random, or somehow constrained by limits of physical possibility—or are they, maybe, directed toward certain purposes by a higher power? If variations are random, then purposefulness (the philosophers of science call it teleology) disappears from the living world. Gone, zero, zip.

Whoa. That's a large step into darkness. No higher purpose

to the vast pageant of life and death? No higher purpose to Herschel's "mystery of mysteries," the first appearance of new species? No higher purpose behind adaptation and diversification, the processes whereby simplicity gives way to complexity on a spectrum from primal ooze to humans? These were implications that Darwin's nineteenth-century audience found difficult to accept. They're still difficult to accept. But this generalized loss of teleology is abstract and impersonal. It's not the source of keenest discomfort with natural selection. Another corollary of the theory is more acutely problematic: loss, for the human species, of our own special status as God's chosen.

Is there a glorious end for which evolution has produced mankind? Are humans in any sense uniquely ordained? Did the deity foresee we were coming and somehow will it? Or are we merely the most well adapted, cerebral, and successful species of primate that has ever lived? Beneath those questions lies one deeper question about the variations from which natural selection has shaped *Homo sapiens*: What is their source?

Darwin proposed in *The Origin* that variations occur in response to "conditions of life"—that is, external stresses such as severity of climate, food shortage, or habitat disturbance, which somehow unsettle the reproductive system. It was a calculated guess. Elsewhere in the book, he admitted: "Our ignorance of the laws of variation is profound. Not in one case out of a hundred can we pretend to assign any reason why this or that part differs, more or less, from the same part in the parents." Scholars have noticed this lacuna: His theory depends on variations, but there's no good account of their origin in *The Origin*. He just didn't know where they came from, nor how. At the time, no one did.

Puzzled as Darwin was about their source, he strongly suggested that variations are, overall, directionless. That is, they go here and there, haphazardly. They are scattershot, not aimed. The point is so crucial, and the language by which he addressed it so tricky, that it deserves a moment's special attention. Back in 1844, in his unpublished 189-page draft of the theory, Darwin had written that variations occur "in no determinate way." Earlier still, in one of his notebook jottings, he had described them as "accidents." He wrote in *The Origin* of "chance" variations, then noted elsewhere in the book that to say they are "due to chance" is an incorrect expression, a convenience of speech that merely "serves to acknowledge plainly our ignorance of the cause" of each one. It's an incorrect expression, he meant, in that variations do have physical *causes*; they just don't have preordained *purposes*. For instance, a drought might increase the rate of variation in a species, he thought, without necessarily evoking any particular variations that improve a creature's tolerance for drought. Or the drought might yield one variation for drought tolerance plus five others that are useless or harmful. If so, natural selection would tend to preserve and multiply that one. Selection is directional; variation, offering raw material to the selection process, is not directional.

But if variations are undirected, and if natural selection calibrates only the fitness of each individual creature to survive and reproduce, then is it possible to believe that God created humans in His image and likeness, endowing us with a spiritual dimension not shared by the best-adapted orchid or barnacle? Arguably not. There's a genuine contradiction here that can't easily be brushed away. But let's be clear: This is not evolution versus God. The *existence* of God—any sort of god, personal or abstract, immanent or distant—is not what Darwin's evolutionary theory challenges. What it challenges is the supposed godliness of Man—the conviction that *we* above all other life forms are spiritually elevated, divinely favored, possessed of an immaterial and immortal essence, such that we have special prospects for eternity, special status in the expectations of God, special rights and responsibilities on Earth. That's where Darwin runs afoul of Christianity, Judaism, Islam, and probably most other religions on the planet.

Victorian scientists such as Adam Sedgwick, the crusty old Cambridge professor who had taught Darwin field geology before the *Beagle* voyage, recognized this challenge clearly, and hated *The Origin* for it. Sedgwick called the book "a dish of rank materialism cleverly cooked and served up." Richard Owen, who had studied gorilla anatomy in his lab, ridiculed Darwin for suggesting that "man might be a transmuted ape." St. George Jackson Mivart, a convert to Catholicism and a former student of Huxley's, became a zealous evolutionist but balked at natural selection and argued trenchantly against it, positing instead some "internal innate force" as the driving cause of evolution. Whatever produced the physical transmutations from one species to another, Mivart added, could never account for the human mind and soul, which existed in a realm untouchable by evolutionary theory. These critics weren't deluded or paranoid about what was at stake. They may have failed to absorb the details of Darwin's theory, and caricatured it in print, but they didn't misconstrue its implications. The denial of humanity's special status, implicit in the idea of natural selection acting on undirected variations, acutely distressed many of Darwin's contemporaries—not just religious leaders and scriptural literalists but also some scientific colleagues, such as the botanist Asa Gray at Harvard, the entomologist Thomas Wollaston, and Darwin's old friend and counselor Charles Lyell. Their distress was well founded. And this was the point, too, over which Emma Darwin quietly suffered forty-five years of philosophical discordance with her adored and adoring husband.

Scientific insight and religious dogma had never come more directly into conflict. It was a bigger issue than whether humans and monkeys share a common ancestry. It was the issue of whether humans and monkeys, along with lobsters and dandelions and all other living creatures, share an absence of special divine appointment. In plain language: a soul or no soul? An afterlife or not? Are humans spiritually immortal in a way that chickens and cows aren't, or just another form of temporarily animated meat?

Today we tend to overlook this horrible challenge implied by Darwin's idea. Theistic evolution has supposedly made the theory safe for people of all faiths. But the deep materialism of Darwin's vision couldn't so easily be overlooked back when natural selection was a shocking novelty. It assaulted sensibilities. It impeded uptake. Most people nowadays aren't aware that, at the time of Darwin's death in 1882, and for two generations afterward, his explanatory mechanism was severely doubted, resisted, and then generally rejected, while evolutionists groped for less repellent alternatives.

37

One of the first serious critiques of Darwin's theory came from William Thomson, the Scottish mathematician and physicist later known as Lord Kelvin. In 1866, Thomson published a short paper based on his calculations of elapsed time since Earth had formed and solidified. Titled "The 'Doctrine of Uniformity' in Geology Briefly Refuted," it was a one-paragraph snort of disdain, playing on the word "Briefly" while asserting that all earthly history was shorter than some persons supposed. Thomson's primary target was Charles Lyell, whose uniformitarian view of geological processes entailed slow, steady action over huge stretches of time; Darwin's concept of slow, steady evolution by natural selection came into question secondarily. Thomson assumed that our planet, having originated as a gob of molten material pulled from the Sun, cooled at a determinable rate as it radiated heat into the chill of space. Given the hot core of magma still remaining, he figured that Earth was probably no more than 100 million years old. That left insufficient time, Thomson argued, for Lyell's pokey gradualism to have accomplished so much geological change. Darwin, whose thinking was grounded in Lyellian geology, felt the pinch too. One hundred million years was far less than the "incomprehensibly vast" amount of time he had posited for natural selection to shape all life as we know it.

The pinch tightened several years later when Thomson, having reworked his numbers and considered other factors, began revising the estimate downward. Make that 30 million years, he said. Or maybe just 10 million. Was it possible to believe that Earth's solid crust was as young as Thomson proposed? Not if you hoped to explain the entire pageant of life—from its Pre-Cambrian doldrums, through the Cambrian explosion of new forms, to the Silurian trilobites, the Devonian ammonoids, the rise and fall of dinosaurs, the age of mammals, and the interesting later trajectory of a certain ape lineage—as resulting from small, undirected variations shaped by natural selection. And vice versa: If you accepted Thomson's clock, you had to reject Darwin's scenario. In 1868, now pressing his case explicitly against Darwin, Thomson told an audience that the limitation of time, although it didn't refute transmutation per se, did seem "sufficient to disprove the doctrine that transmutation has taken place through 'descent with modification by natural selection.'"

Darwin grumbled to Alfred Wallace about the "odious spectre" of Thomson, and in revising *The Origin* for its fifth edition he cut "incomprehensibly vast" to merely "vast," a reluctant fillip of compromise. He also inserted several sentences acknowledging the difficulty of measuring earthly time, and he conceded that "we have no means of determining how long a period it takes to modify a species." Darwin's confidence was tested, but not broken. He could make adjustments.

Another negative commentary came from Fleeming Jenkin, a professor of engineering who would later become a business partner of Thomson. Jenkin's long review of *The Origin*, carried by *The North British Review* in 1867, criticized Darwin for several supposed mistakes of logic and judgment—most notably, one involving inheritance. Jenkin's assumption, not unusual for his time, was that the mixing of bloodlines in sexual reproduction brings a proportionate mixing of attributes. If a white man mates with a black woman, the children will be mulatto. If a long-necked goose breeds with a short-necked goose, the goslings will be medium-necked. If a white-flowered plant is crossed with a red-flowered variant, their offspring will flower in pink. True? Not necessarily. Nowadays this is known as "blending inheritance," a spurious simplification of what actually occurs. But blending inheritance was the sensible-sounding premise from which Jenkin argued, and Darwin had no better theory of inheritance with which to answer him.

Such blending, Jenkin tried to show, was fatal to Darwin's theory. Granted, small beneficial variations might increase the reproductive success of some individuals. But in the process of interbreeding, Jenkin thought, those variations won't be passed along intact. They'll be diluted by half in each new generation (assuming that only one parent carries the novel trait), and therefore eventually blended away to nothingness. "Fleeming Jenkin has given me much trouble," Darwin confided to Hooker, around the time he finished work on that fifth edition. He had anticipated the blending inheritance problem himself, as far back as 1838, in his transmutation notebook "C," when he ruminated vaguely about "the tendency to revert to parent forms." Now Darwin dealt with Jenkin's objection, as best he could, by emphasizing a distinction between single variations that appear rarely in a population and variations that appear in numerous individuals simultaneously. The latter sort, allowing a reasonable possibility for two variant individuals to breed with each other, wouldn't so easily be blended away.

It was an emergency hedge, and not very persuasive. There was a much better answer to Jenkin's criticism, but that would become available only later, with the rediscovery of work by Gregor Mendel.

The unkindest cut against natural selection was delivered by Alfred Wallace, of all people, more than a decade after their joint publication. By then Wallace had been home from the East for seven years and written a magnificent book of travel reportage and natural history, *The Malay Archipelago* (published in 1869). He had also solidified his friendship with Darwin. He never became one of Darwin's close confidants, like Hooker or Fox, but he was a very special colleague: the co-discoverer and co-defender of the notorious theory. Apart from Darwin himself, no one understood natural selection better or applied it more forcefully than Wallace. His zealotry, in fact, sometimes surpassed even Darwin's. Wallace saw natural selection operating in certain cases—the gaudy plumage of male pheasants, for instance—where Darwin favored a different causal mechanism. (Darwin's alternative was *sexual* selection, the idea that runaway preferences by the opposite sex, not imperatives of survival, are what drive such gratuitously elaborate modifications.) Despite his strong intellectual commitment to the theory, Wallace evidently felt no urgency about asserting his own claim of authorship. He barely mentioned natural selection in *The Malay Archipelago*, and then with modesty approaching coyness, as an idea "elaborated by Mr. Darwin in his celebrated *Origin of Species.*" A year later, he reprinted his Linnean Society paper, as well as the "law" paper of 1855 and several others, in a volume he called *Contributions to the Theory of Natural Selection*, of which the title seems to reflect how he viewed himself: as a contributor to Darwin's theoretical breakthrough. Wallace waited until 1889 to produce a full-length volume on the subject, that one abnegatingly titled *Darwinism: An Exposition of the Theory of Natural Selection with Some of its Applications.* He was an independent-spirited man but, excepting a few circumstances, he remained faithfully subordinate to Darwin and Darwin's idea. The most notable exception came in early 1869, when Wallace unexpectedly dissented on a crucial point, asserting that natural selection couldn't account for the human brain.

Wallace's apostasy may have reflected other changes in his own life and interests since returning to England. Always eclectic, impetuous in his enthusiasms, he had gotten interested in spiritualism and begun attending séances as an avid believer. During one spooky session with a medium, he'd heard his dead brother Herbert rap out a coded hello from the beyond. Spiritualism was enjoying a boom in popularity at the time, thanks presumably to its combination of vulgar metaphysics, nostalgia for the dearly departed, and parlor entertainment in an era before television. Some scientists saw it as a harmless fad, or else as contemptible bunkum, but to Alfred Wallace it was a new horizon in anthropology. Although not religious in any conventional sense, he concluded that there was more to this world than physical causes and effects. His new credence didn't collide openly with his older views until April 1869, when *The Quarterly Review* carried an essay by him, mostly focused again on Lyell's geology, in which Wallace pointedly digressed to the subject of natural selection. That mechanism couldn't have produced the human brain, he wrote, let alone the "moral and higher intellectual nature of man." Of course the living world is governed by laws, Wallace noted. But he himself was now inclined to believe "that an Overruling Intelligence has watched over the action of those laws, so directing variations and so determining their accumulation" as to yield the loftier, more wondrous human capacities.

Darwin knew the article was coming and, a month earlier, had told Wallace with nervous good cheer: "I shall be intensely curious to read the *Quarterly*. I hope you have not murdered too completely your own & my child." In the event, it was as bad as he feared: intellectual infanticide. Natural selection as Darwin (and Wallace too?) had originally conceived it was meaningless if "an Overruling Intelligence" overruled the haphazardness of the variations, directing them toward foreordained purposes. In the margin of his copy, Darwin scratched "No!!!"

38

Discomfort with natural selection, working crosswise to the general acceptance of evolution, pushed biologists during the late nineteenth century toward alternate explanatory mechanisms. Some of those biologists looked backward in time, and to France, for a revival of Lamarckism. Some embraced other evolutionary theories, various in their particulars but with enough common elements to be lumped under two labels, "orthogenesis" and "saltationism." All three—orthogenesis, saltationism, and revived Lamarckism—came on strongly during the 1880s and 1890s, a slump period for Darwin's reputation. A historian of evolutionary thought named Peter J. Bowler has nicely charted these currents in several of his books, including one titled *The Eclipse of Darwinism*. Bowler's research corrects the misconception that, after publishing *The Origin of Species*, Charles Darwin rode to glory in a flaming chariot. No, he sat parked on a siding.

The new Lamarckians didn't entirely reject Darwin's big idea, but they considered it small. Okay, they allowed, maybe natural selection does play a marginal role in fine-tuning adaptations, but it can't explain the origin of variations or the dramatic trends and patterns of evolutionary change. Selective in their definition of Lamarckism, they also largely ignored Lamarck's own woozy notions about "subtle fluids" and a "feeling of existence," preferring two other items from his cornucopia of theory: the parallel progression of independent lineages from simple to complex (that is, the prairie-grass model of biological diversity, as opposed to the branching-tree model) and the inheritance of acquired characteristics. They emphasized the role of environmental conditions in eliciting need-directed variations (not undirected ones, as Darwin had it) which, so they believed, are heritable. They also inclined toward the view that long-term evolutionary trends are linear, triggered by environmental conditions and driven onward by habit, and by the inheritance of what habit produces. Horns get bigger, from species to species over millions of years, because animals use them when they smack heads. The fossil record yields other examples of such linearity, supposedly transcending the immediate adaptive needs of each individual creature and expressing inherent trends throughout the history of a lineage. Direct influences from the environment might account for the small-scale changes and adaptations, while continuous habit or some mysterious force drives the long-term trends.

There was no perfect consensus, though, among the new Lamarckians. Paleontologists tended to see the long, linear trends; field naturalists and lab experimenters tended to see, or imagine they saw, the inheritance of acquired characteristics. The whole school of thought was especially strong in America, where a naturalist named Alpheus S. Packard, Jr., called it neo-Lamarckism.

Packard, like other influential neo-Lamarckians of his generation, had been trained by Louis Agassiz, the Swiss-born naturalist who ruled in professorial majesty at Harvard's Museum of Comparative Zoology. Agassiz was a brilliant but obdurate man, an essentialist who detested evolutionism—Darwin's brand in particular—and clung to a vision of well-ordered nature assembled by special creations. The zoology of Agassiz was consonant with the natural theology of William Paley. Agassiz's livelier students, such as Packard, did cross the line to accept evolution in principle, but even they mostly retained the old man's distaste for Darwin's cold, hard mechanism. Packard began seeing what he thought were Lamarckian phenomena during his study of the horseshoe crab, *Limulus polyphemus.* Then he turned to the blind insects and other dark-dwelling animals of Mammoth Cave, in Kentucky, concluding that their loss of eyesight (in some cases, of eyes altogether) results from disuse, followed by shrinkage of visual organs, followed by inheritance of the shrunken forms. Although Darwin himself had admitted a secondary role for use and disuse, the evidence from Mammoth Cave struck Packard as "Lamarckism in a modern form." That explanation seemed to him "nearer the truth than Darwinism proper or natural selection."

Among the American paleontologists was another Alpheus (you can't tell them without a scorecard), also trained at Harvard by Louis Agassiz: Alpheus Hyatt. From his study of ammonoids and other fossil invertebrates, Hyatt concluded that evolution is an additive developmental process—one that proceeds by the addition of new adult characteristics onto older sequences of development. Adding such characteristics, Hyatt thought, somehow compresses the more primitive traits backward into earlier embryonic stages. This idea became known as "the law of acceleration," suggesting that speedier growth through the earlier stages allows for the additional complexity in adulthood. What's the source of those newer, more complex characteristics? After some hesitation, Hyatt accepted the Lamarckian view that they are adaptive adjustments to environmental stresses, acquired as habits and then inherited.

Edward Drinker Cope, an American paleontologist working on vertebrate fossils, arrived at the law of acceleration independently. Like Hyatt, he saw long-term linear trends in fossil sequences—new modifications added onto older forms in a steady, directional way—and, again like Hyatt, became convinced that the inheritance of characteristics acquired in response to environmental conditions is what best explains them. In 1877, Cope published a book, *The Origin of the Fittest*, combining Darwin's own title with Herbert Spencer's vivid phrase ("the Survival of the Fittest") to impute that Darwin hadn't dug deeply enough into the subject. Like the other Alpheus—that is, Alpheus Packard—Cope granted that natural selection might play some part in culling inferior individuals, but he figured it couldn't be more than secondarily important, because it didn't explain the source of variation. Lamarckism, Cope thought, did.

In England, Herbert Spencer himself had espoused a theory of evolution (he called it "the development hypothesis") as early as 1852, seven years before Darwin's book. Spencer was no biologist; he worked as a journalist, became prominent as a pop philosopher, and picked up his evolutionary ideas from reading Lyell's dismissive account of Lamarckism (which turned him, perversely, toward Lamarck) and the mystery bestseller of the day, *Vestiges*. His own writings about evolution were grandiose and murky, detached from the sort of empirical detail that Darwin offered abundantly. But the subject gave Spencer's works on political philosophy and sociology a yeasty fluff, especially when he linked his advocacy of laissez-faire individualism to notions of evolutionary progression; and he sold well. Some scholars credit (or blame) Spencer for launching the intellectual movement misleadingly known as Social Darwinism, and for transmitting it to America through his publications and, more personally, during a visit in 1882. The steel magnate Andrew Carnegie read both Spencer and Darwin, finding luminous reassurance in being told that harsh competition is a constructive law of nature. By then Spencer himself had emerged as a neo-Lamarckian, and arguably a social neo-Lamarckian, too. In a choice between natural selection of undirected variations, on the one hand, and heritable advantages gained through striving, on the other hand, the latter fit better with his ideas about self-advancement. Onward and upward for ambitious men and their scions! Eleven years after Darwin's death, Spencer made it explicit with an essay titled "The Inadequacy of Natural Selection."

The list of notable neo-Lamarckians in Britain and Europe also included Arthur Dendy (a paleontologist), Samuel Butler (a novelist and argumentative proselytizer), George Henslow (a clergyman-naturalist, who wrote a book about the "self-adaptation" of plants to their living conditions), Joseph T. Cunningham (a marine biologist, who studied color change in flatfish), Peter Kropotkin (a Russian aristocrat turned socialist, who argued that cooperation among animals, as a heritable habit, might be more important than natural selection), C. E. Brown-Séquard (known for his experiments inducing heritable epilepsy in guinea pigs), and the zoologist

Theodor Eimer. By the end of the 1880s, as Samuel Butler gloated, nearly every issue of the journal *Nature* (founded by Darwin allies in 1869) contained something on Lamarckian inheritance.

Theodor Eimer, professor of zoology at Tübingen, Germany, was an important transitional figure between neo-Lamarckism and another non-Darwinian school of thought, for which Eimer himself popularized the label "orthogenesis." Early in his career, Eimer studied wall-climbing *Lacerta* lizards on the island of Capri. Later he investigated the color patterns of butterfly wings. In the first of his two major volumes on evolution, published in 1888 as *Entstehung der Arten* (with an English edition soon afterward, translated as *Organic Evolution*), he combined a Lamarckian view of character acquisition with a claim that internal "laws of growth" dictate the characteristics to be acquired and, over the long term, the direction in which evolution goes. For certain traits, the direction might be neutral—or worse—with regard to adaptation. The word "orthogenesis" means growth in a straight line. It implies an inherent tendency of some sort, expressed ever more extremely in one descendant after another and independent of the creatures' immediate needs. This view became popular among paleontologists (including Cope and Hyatt in America) as an explanation of certain linear trends in the fossil record, some of which appeared not just non-adaptive but destructive. The Irish elk, *Megaloceros giganteus*, is one famous example of what supposedly can result from orthogenesis; its antlers grew so oversized that they seemed to have doomed the species to extinction. Eimer saw similar phenomena, he thought, among butterflies. His studies of Lepidoptera convinced him, says Peter Bowler, "that the actual course of orthogenetic evolution was completely predetermined by the internal predisposition to vary in a particular direction."

What accounts for the "internal predisposition"? Neither Eimer nor Hyatt nor Cope nor anyone else ever offered a mechanism to account for how this amazing process might work. But it seemed to offer them some satisfaction that Darwin didn't. Eimer's second big volume appeared in 1897, just before he died, under a wonderfully tongue-twisting German title, *Orthogenesis der Schmetterlinge: ein Beweis bestimmt gerichteter Entwickelung und Ohnmacht der Natürlichen Zuchtwahl bei der Artbildung*, which is translatable as *Orthogenesis of Butterflies: A Proof of Definitely Directed Development and the Weakness of Natural Selection in the Origin of Species.*

A person might well ask: If it's Definitely Directed Development, then directed *by what*? Not by God, so far as Theodor Eimer and other orthogenesists were concerned, and not by the imperatives of adaptation.

Saltationism embodied the view that evolution proceeds by leaps. Darwin had explicitly rejected that idea in *The Origin*, citing what he considered a reliable old maxim: *Natura non facit saltum.* It was true, he wrote, that nature makes no leaps, because natural selection "must advance by the shortest and slowest steps." Huxley had disagreed, believing that nature does indeed move by smallish jumps, and he worried that Darwin had burdened his theory with an unnecessary difficulty. During the late 1880s, a British zoologist named William Bateson came to share Huxley's dissatisfaction with Darwin's gradualism, especially after tossing aside his laboratory approach in favor of fieldwork on the steppes of central Asia. Since species are discontinuous, one from another, Bateson argued, the variations from which species are produced might be discontinuous, too. He went still further: Discontinuous variation *is* evolution. Natural selection isn't necessary, Bateson thought, if variation occurs in big, sudden leaps that sometimes yield new species. The Dutch botanist Hugo De Vries reached the same conclusion about the same time, based on his study of discontinuous variations in the evening primrose, *Oenothera lamarckiana*. De Vries put an old word to new use for such sudden, major changes, calling them "mutations."

By the end of the 1890s, natural selection as Darwin had defined it—that is, differential reproductive success resulting from small, undirected variations and serving as the chief mechanism of adaptation and divergence—was considered by many evolutionary biologists to have been a wrong guess. It was interesting in its historical context, they conceded, as the pet idea of the man who had opened the world's eyes to evolution. Possibly it did play some small, secondary role. Or possibly none. There were too many damning arguments against it, such as Jenkin's about blending inheritance and Thomson's about planetary age. There were too many newer ideas, such as saltationism, and older ones, such as Lamarckism, that carried stronger intuitive appeal.

But something was missing from all the alternate theories, as it was missing from Darwin's own: a clear understanding of how inheritance works. During the last years of the century, for one instance, Hugo De Vries began writing his evolutionary opus, *Die Mutationstheorie*, with bold notions about the abrupt origin of new species but little appreciation for the routine dynamics of heredity and incremental change. When his first volume was nearly complete, a colleague sent him a small packet, with a note: "I know that you are studying hybrids, so perhaps the enclosed reprint of the year 1865 by a certain Mendel, which I happen to possess, is still of some interest to you." Turns out, it was of interest to everybody.

39

One of Darwin's great strengths as a scientist was also, in some ways, a disadvantage: his extraordinary breadth of curiosity. From his study at Down House he ranged widely and greedily, in his constant search for data, across distances (by letter) and scientific fields. He read eclectically and kept notes like a pack rat. Over the years he collected an enormous quantity of interconnected facts. He looked for patterns among those facts but was intrigued equally by exceptions to the patterns, and by exceptions to the exceptions. He tested his ideas against complicated groups of organisms with complicated stories, such as the barnacles, the orchids, the social insects, the primroses, and the hominids. Gregor Mendel was a different sort of scientist, with a different cast of mind. He lived in a monastery and studied peas.

The monastery was an Augustinian one, in Brno, an ancient town southeast of Prague in what was then part of greater Austria. Mendel's experimental organisms were the common garden pea, *Pisum sativum*, and its close relatives. Luckily for him, the genetics of *Pisum* happen to be simple and straightforward in a way that those of *Oenothera lamarckiana* and many other organisms are not. After eight years of crossbreeding experiments, in which he tracked the inheritance of traits in flower color, leaf size, stalk length, seed shape, and other easily visible aspects of a pea plant, Mendel described his work to fellow members of the Brno Natural

History Society. That was early in 1865. His results included several important observations: that some traits are *dominant*, whereas others are *recessive* (Mendel's terminology, adopted from earlier workers); that a dominant trait, when crossed with a recessive, is transmitted intact to the next generation, not in diluted or compromised form; that a recessive trait becomes latent when crossed with a dominant, but appears in full force when crossed with a similar recessive; and that, after a large number of crossings between any pair of dominant and recessive traits, the ratio among offspring will be almost exactly 3 to 1. Crossing red-flowered plants with white flowered plants, for instance, Mendel got 705 red-flowered offspring and 224 white-flowered, for a ratio of 3.15 to 1. Crossing puffy-pod plants with constricted-pod plants, he got a ratio of 2.95 to 1. Round-seeded plants crossed with wrinkle-seeded plants gave him 2.96 to 1. The average from seven experiments was an overall ratio of 2.98 to 1, registering a mystifying consistency that couldn't be coincidence.

The implications were huge. With these experiments, Mendel had shown that heredity functions by way of indivisible, particulate units, only two in each case, and not (as Darwin and others believed) by way of cumulative masses of tiny elements afloat in the blood. He had demonstrated that each parent contributes just one hereditary particle, not a profusion of them, for any given trait. His 3-to-1 ratio reflects the four different ways that two parental particles can be combined in a second-generation individual, given that each parent might contribute either a dominant particle (call it *A*) or a recessive particle (call it *a*) for the given trait: *AA, aa, Aa, aA*. Of those four possibilities, three (*AA, Aa, aA*) will result in manifestation of the dominant trait, while only one (*aa*) will produce the recessive trait. Mendel had outlined a central law of heredity and pointed toward the concept of the gene. He had also suggested the modern distinction between phenotype (what the organism shows) and genotype (what the organism carries). He had punctured the illusion of blending inheritance.

Like the Darwin-Wallace presentation to the Linnean Society, Mendel's lectures made no great impression at the time. A year later, published in the Brno Natural History Society's journal under the modest title "Experiments in Plant Hybridization," they made no great impression again. Mendel himself arranged for about forty reprints to be sent to botanists and other scientists who might have been interested; but the interest didn't flare. His paper lay almost entirely unnoticed and unused for thirty-four years. Why? Was he too far ahead of his time? Yes, in the sense that he offered answers to questions that hadn't yet been clearly enough asked. Was he ignored by the scientific community because of his monastic isolation and obscurity? Yes, that didn't help either—Brno wasn't London, and its Natural History Society was an improbable venue for announcing a major scientific breakthrough. Was he disadvantaged by the fact of having published only one notable paper, not a body of interrelated work? Somewhat. There's no single reason, just a handful of contributing factors, to account for this accident of neglect. You could say that Gregor Mendel was too modest and unassuming to call attention to himself. That he was unlucky. That biology itself was unlucky. That he made the fatal mistake, for his follow-up studies, of shifting from peas to a more complicated group of plants, the hawkweeds. That he got distracted from further plant experiments by his election as abbot of the monastery. Anyway, for all the response Mendel's article evoked, at least during his lifetime, he might as well have buried his forty reprints in the garden. Then, in 1899, a copy of his paper was mailed to Hugo De Vries. It may have been one of those original forty that Mendel himself had hopefully cast forth.

In the meantime, a German zoologist named August Weismann had developed his own theory of inheritance, containing several strong ideas. One of those ideas was that heritable traits pass from generation to generation by way of molecular material contained in the nuclei of cells. A second was that, contrary to Lamarckian and neo-Lamarckian belief (including Darwin's own Lamarckian misconception), acquired characteristics are not inherited. Never. In no cases. Not possible, according to Weismann. He argued that the *germ plasm* (the cell line that eventually produces gametes, reproductive cells, such as eggs and sperm) stands isolated from the *soma* (the rest of the body), and that it can't be altered within an individual by neck-stretching, weightlifting, blacksmithing, cave dwelling, drought, severe cold, or any other activities or environmental conditions affecting the body. The soma is what changes with habit or stress, Weismann argued; the germ plasm remains untouched; and changes to the soma aren't heritable. More clearly than Mendel (without having read about Mendel's peas), he saw the distinction between genotype and phenotype. Building on recent insights in cell biology, he also recognized another important phenomenon: that the haphazard intercrossing of chromosome branches, during the process of cell division to form gametes, results in chromosomal recombination. That is, tangling, breaking, and reconnecting. In sexual reproduction, this intercrossing continually generates a richness of possible combinations, and therefore an abundance of variations among offspring—even among offspring of the same parents. Biologists today realize that such recombination of existing genes, along with outright mistakes of gene duplication that create wholly new gene forms (now known by De Vries's term, mutations), are the main answer to the lingering question that clouded Darwin's work and his successors' for decades: *What is the source of variation?* Mutation and recombination supply most of it.

Mutation produces new variants of existing genes. Recombination generates variation by splicing together new gene combinations from one chromosome to another. In the process of meiosis (double cell division, producing gametes), normal and mutant genes on their normal or recombined chromosomes are parsed out into the reproductive cells. This egg gets an *A*, plus *BCdEF*. That egg gets an *a*, plus *BCDEf*. Another egg gets a *mutant-a*, plus *bcdeF*. Shuffle, cut the deck, add some jokers, shuffle again. Insofar as mutation and recombination are accidental processes, variation is undirected by need or purpose. Natural selection acts upon it. Mendelian inheritance prevents the results from being blended away.

40

It's not my intention, so near the end of this little book, to try to quick-step you through all the major episodes in the later history of evolutionary biology. That would take me too far beyond my allotted length and too far out of my depth.

If it *were* my intention, I'd be obliged to describe how the saltationists seized on Mendel's rediscovered ideas, thinking that particulate inheritance supported their arguments against natural selection, but they were wrong; how Weismann's concept of the isolated germ plasm led to a strict view of natural selection as the sole evolutionary mechanism—a view more Darwinian than Darwin himself—that came to be known as neo-Darwinism; also how Thomas Hunt Morgan's research on fruit fly genetics and Richard Goldschmidt's notion of the lucky mutant (or, as he called it, "the hopeful monster") brought saltationist thinking well into the twentieth century; and how saltationism eventually faltered and dissolved in the face of brilliant new work in mathematical genetics, mostly by R. A. Fisher, J. B. S. Haldane, and Sewall Wright, showing that Mendel's particulate inheritance actually supported Darwin's selection theory rather than confuting it. Having mentioned Sewall Wright, I'd offer at least a parenthetical explanation of his concept of *genetic drift*, a random process that becomes very important in small, isolated populations and (as some biologists believe) may be largely responsible for speciation events. I would also remind you that the discovery of radioactivity by Henri Becquerel, at the end of the nineteenth century, furnished a decisive rebuttal to William Thomson's cavils about the age of planet Earth (its internal heat source now better understood) and recast the estimates of elapsed time, allowing Darwin all the eons necessary for evolution by natural selection. Most important, I would need to outline an intellectual event known as the Modern Synthesis, during the 1930s and early '40s, in the course of which George Gaylord Simpson (a paleontologist), Theodosius Dobzhansky (a geneticist), Julian Huxley (a wide-ranging biological thinker, grandson of Darwin's friend T. H. Huxley), Ernst Mayr (a naturalist and systematicist), and several other influential biologists, building on the work of Fisher, Haldane, and Wright, unified Mendelian genetics with Darwinian selection and established a synthesized theory of evolution, roughly as it's accepted today. I say "roughly," of course, because even their Modern Synthesis is no longer modern. In the past sixty years it too has been critiqued, modified, added to, and otherwise improved. I would be duty-bound, in addition, to touch on some latter-day developments and modifications, such as Ernst Mayr's hypothesis about genetic revolutions among insularized populations, Niles Eldredge and Stephen Jay Gould's concept of punctuated equilibria, Motoo Kimura's neutral theory of molecular evolution (along with Richard Lewontin's response to it), the thinking of George C. Williams and Richard Dawkins on selfish genes, Edward O. Wilson's provocative overview of sociobiology, Stuart Kauffman's fascinating suggestions about self-organization emerging from complex genetic systems, and much more. Whew. But no, I won't try to do all that. Not here, not now.

For lucid accounts of those developments, if you happen to want them, you can turn to sources such as Ernst Mayr's readable (but not disinterested) history, *The Growth of Biological Thought*; or Peter J. Bowler's various books, including *The Non-Darwinian Revolution*; or Douglas Futuyma's excellent textbook, *Evolutionary Biology*; or Mark Ridley's, *Evolution*; or David J. Depew and Bruce H. Weber's dense survey, *Darwinism Evolving: Systems Dynamics and the Genealogy of Natural Selection*; or Stephen Jay Gould's ponderous but richly informative (it should be, at 1,433 pages) intellectual doorstop, *The Structure of Evolutionary Theory*; or . . . a passel of other books, some good and some merely useful. Darwin's theory, as I warned you at the start, has attracted a vast amount of scholarly nibbling and scribbling. But the fascinations and the implications of that theory are vast, too. And the story hasn't ended. It continues to unfold.

The central themes of the story, as told by Mayr or Gould or most of those others, are that evolution is real and wondrous, and that the idea of natural selection has survived and succeeded because it fits the observable facts better than any alternative idea, doing exactly what a scientific theory must do: explain material effects by way of material causes. As Darwin himself conceded—and as Sewall Wright, Motoo Kimura, and certain other biologists have since affirmed—natural selection isn't the sole mechanism of evolutionary change. But it's the primary mechanism. It's the lathe and the chisel that shape adaptations. It's the central concept of Darwinism, whatever else Darwinism might be taken to include. It's the starting point for understanding how evolution works.

When first published in *The Origin of Species*, according to Douglas Futuyma's textbook, Darwin's long argument was "based on logic and on interpretation of many kinds of circumstantial evidence, but he had no direct evidence." Biogeography, paleontology, embryology, morphology—those could all be considered indirect, in that their puzzling patterns were explicable by Darwin's theory. More than seventy years had to pass before a synthesized understanding of Mendelian heredity and Darwinian selection would, in Futuyma's words, "fully vindicate his hypothesis." But the vindication came. In 1959, the centennial of *The Origin* was celebrated in a spirit of confidence that Darwin, the crafty old boy, had gotten it right. Later discoveries have added further certainty. And more every year.

Not long ago I visited Futuyma in his office at the University of Michigan. A long narrow table, down the middle of a long narrow room, was strewn with journal papers. His shelves were full of books. No fruit flies in cages, no ammonoid fossils, no pickled barnacles. It was a place for thinking and chatting. Futuyma is a mild, urbane, and very smart man with short-clipped gray hair and wire-rim glasses. On this day he was wearing a bulky sweater. I had come to ask him about the evidence for evolution.

In response to my questions, he moved quickly through some familiar points—vestigial organs, the fossil record, patterns of biogeography—and talked mostly about molecular genetics. He reminded me that molecular biologists generally haven't concerned themselves with the same questions, let alone the same answers, that engage evolutionary biologists. For fifty years, since Watson and Crick discovered the structure of DNA, the molecular people have been interested in genes, proteins, and the ways they function within living cells, but not much in species and the ways they evolve. At the University of Michigan and many other universities, the two disciplines—molecular biology and evolutionary biology—don't even reside within one department. That said, Futuyma pulled out his heavily marked copy of *Nature* for February 15, 2001. It was a historic issue, fat with articles on the results of the Human Genome Project. Beside it he slapped down a more recent issue of *Nature*, also thick and important, this one devoted to the sequenced genome of the common mouse, *Mus musculus*. The particular strain of mouse under scrutiny was known as C57BL/6J, a laboratory lineage frequently used in research. The headline of the lead editorial announced: HUMAN BIOLOGY BY PROXY.

The mouse genome effort, according to *Nature*'s editors, had revealed "about 30,000 genes, with 99% having direct counterparts in humans." What they meant by "direct counterparts" was not identical genes (like the many identical genes that humans and chimpanzees share) but very similar ones. Such a high degree of similarity is nevertheless dramatic. Mice and humans possess about the same number of genes, almost all of them direct counterparts, "and we both like cheese," *Nature* noted. "So why aren't mice more like us? The answer probably lies in the regulation of those genes." Similar genes produce humans on the one hand, mice on the other hand, because of how they are turned on and off during each creature's embryonic development and growth.

Futuyma helped me understand this in a broader context. The resemblance between our 30,000 human genes and the 30,000 mousey counterparts, he said, represents another form of homology, like the resemblance between a five-fingered hand and a five-toed paw. Now consider: Would any wise and busy God fabricate a human species, by special creation, that is similar to mice in 30,000 ways? Not likely. In fact, not rationally imaginable. Homology so intricate can be explained only by common descent. Flipping to the end of the main article, Futuyma read me a sentence: "Comparative genome analysis is perhaps the most powerful tool for understanding biological function." He looked up. "That's a strong statement from molecular biologists." Reading again: "Its power lies in the fact that evolution's crucible is a far more sensitive instrument than any other available to modern experimental science." Evolution's crucible? In plain words: natural selection, preserving genes and discarding them, sometimes one by one.

Futuyma's point was that, after decades of increasing tension between two competing disciplines, even molecular biologists are now beginning to concede that *all* biology is evolutionary biology. "This is the future," he said, "of biological and biomedical science."

The Last Beetle

1876–1882

41

In his later years, Darwin's health improved, but he grew weary.

The old urgency was gone. He knew that the big works of his life were finished. Maybe that's why he vomited less frequently and suffered fewer attacks of bad, dizzy head. He resigned himself grumpily to the inconveniences of fame—worshipful visitors, letters from strangers, requests for his presence or his opinion or his expert testimony in court—though he would still claim incapacity by reason of invalidism when it suited him. For instance, he declined an invitation to go to Oxford for an honorary doctorate. Who needed that? Oxford was full of religious zealots such as John Henry Newman, and besides, Darwin himself was a Cambridge man. He declined to serve as a pallbearer for Charles Lyell, one of his oldest and most supportive friends, when Lyell was buried with great ceremony in Westminster Abbey. Darwin didn't even go up to London for the funeral. He had skipped other funerals and deathbed courtesies over the decades, putting his own needs for privacy and tranquility ahead of his human loyalties, but missing Lyell's interment in 1875 was a clear signal of his increasing disengagement from any community larger than family and village. When his brother Erasmus died several years later, after a lifetime of bachelor schmoozing in London, Darwin had the body brought to Downe for burial in the local churchyard. He probably assumed (but wrongly) that he would eventually lie there himself, elbow to elbow with Emma when she too passed, and not far from his lonely older brother.

Darwin was a selfish and ruthless man in some ways, but selfish and ruthless mainly in service to his work. He was also sweet-spirited and dutiful, with a strong sense of personal morality grounded only in his materialistic notions of how human social behavior had evolved. Occasionally he performed acts of quiet generosity, helping some worthy fellow get a job or a government pension, or sending a sizable check for a good cause. Near the end of his life, he still functioned as treasurer of the Downe Friendly Society, the cooperative savings-and-insurance club that he had helped found for the working people of the village. He also sat on the local school board for some years, and served as a magistrate, adjudicating small cases.

Of the letters that reached him from everywhere, some were weird or peremptory. Dear Mr. Darwin, what are your religious views? Dear Mr. Darwin, I'm trapped in a lunatic asylum, please get me out. Dear Mr. Darwin, I have two alligators in a mill pond here in Yorkshire; what would you like to know about them? He answered many of those letters, and usually with good grace. More than ever, he connected with the world only by publishing books and through the mail.

He enjoyed the small pleasures, distanced and non-disruptive, of being elected to national academies in Hungary, Russia, and the Netherlands, and of receiving (in absentia) a royal order, *Pour le Mérite*, from the king of Prussia. Karl Marx sent him a complimentary copy of *Das Kapital*, with salutations from a "sincere admirer." By some combination of neglect and prudence on the part of successive prime ministers, in consultation with Queen Victoria, Darwin was never offered a knighthood, despite his international renown. (After his death, when it was too late, the government made gestures of rectification: Two of his sons were knighted for lesser achievements.) Pestered by his family, he allowed his portrait to be painted in oils, several times, and he posed for photographs looking patriarchal. People wanted a visual image of Britain's foremost living scientist—whether they had read his books, whether they understood and accepted his ideas, or not. He was a cultural icon by then, a celebrity, mildly notorious as celebrities should be. The most memorable of the photos were several shots taken by an unidentified man from the Elliott & Fry photographic company, who came down from London and captured Darwin, on the veranda, dressed for a bad-weather stroll around the Sandwalk. This was about a year before he died.

You can see his tired, chilly detachment in those photos today. He wears a black cape, pinched tightly around, and a black felt hat, like a bowler with a generous brim. His hands are invisible. His beard is white and shaggy, merging through unruly sideburns with his hair, which is scruffy down the back. His eyes are knowing and glum.

He grew weary, in particular, of his efforts to advance and defend evolutionary thinking, with all its subsidiary theories and ramifications. Natural selection was just part of that, albeit a central part. Under attack by Fleeming Jenkin, William Thomson, and others, Darwin had trimmed his claims about natural selection in successive revisions of *The Origin*, giving more emphasis to Lamarckian use-and-disuse, and to the direct action of external conditions. He never abandoned his brave, scary idea, but he hedged it dispiritedly; in 1880, he wrote to the editor of *Nature*, responding to one more critique with the indignant rejoinder that he had never claimed evolution depends *only* on natural selection. It was true, he never had, not even in his first edition of *The Origin*. But insisting on that point was a sad and unnecessary comedown. He didn't know that Mendel's insights, and radioactivity, and other discoveries would later vindicate his earliest, strongest assertions about natural selection.

Other portions of his later work, from the twenty years of continued productivity after *The Origin* first appeared, addressed hard questions that were still being contested now as Darwin himself faded away. There was his (incorrect) theory of inheritance, as articulated in *The Variation of Animals and Plants under Domestication*. He imagined a process, which he labeled "pangenesis," whereby millions of tiny particles zoom through the body and carry heritable traits, quantitatively, into descendants. There was his (valuable) concept of sexual selection, as propounded in *The Descent of Man*. There were his ideas about the sources of variation, the importance of cross-fertilization as opposed to self-fertilization in hermaphroditic plants, the evolution of moral instincts in humans, and much more.

Although he still cared about those subjects, he found himself unable or unwilling to argue them every time some energized crank dropped a gauntlet. To a man who wrote to him with probing assertions about human behavior, Darwin pleaded that in recent years he'd been working only on plant physiology, and that all other subjects had slipped from his head. It fatigued him, he admitted, to try to bring them back. He thanked another man for his "interesting letter" about hairiness on the ears of infant humans and monkeys, then confessed: "I am so old that I am not likely ever again to write on general and difficult points in the theory of Evolution." It was a polite way of saying: *Leave me the hell alone.* And he was always polite.

The work on plants was quieter, more soothing, less intricately conceptual, and less inflammatory. Some of it carried evolutionary implications—regarding, for instance, the ways by which plants reproduce, acquire variations, and adapt. But writing about heterostyled dimorphic primroses (never mind what those are) didn't seem nearly so provocative as writing about the *os coccyx*, that rudimentary hint of a human tail, as he had done in *The Descent of Man*. By now he'd had a bellyful of being provocative, which brought stress and provocation in return. With help from his son Francis, who had finished Cambridge and gravitated back to Downe village, he continued his ingenuous style of botanical experimentation, keeping sundews and Venus's-flytraps in pots, feeding them insects and raw meat, tormenting them with salts of ammonia to test the sensitivity of their leaves. This led to his book *Insectivorous Plants*, published in 1875. Francis and another son, George, helped with the illustrations. Despite the wonderful luridness of the subject, flesh-eating vegetation, Darwin's treatment was sober and technical, and *Insectivorous Plants* didn't sell so well as his evolutionary volumes. He wasn't deterred. He liked messing around in the greenhouse and the garden. In his study, the potted creatures were good company.

Also in 1875, he brought out *The Movements and Habits of Climbing Plants*, a commercial (but not very) edition of a long paper he'd published ten years earlier through the Linnean Society. John Murray was still glad to be his publisher for general interest books, even when the topics seemed narrow and the sales potential limited, as they increasingly did. A year later came *The Effects of Cross and Self Fertilisation in the Vegetable Kingdom*, which Darwin considered a sort of companion piece to his earlier book on the fertilization of orchids. Within the next several years he produced a new edition of *Orchids* and two other plant books, both containing small insights and discoveries in which he took special pride; but those books were little noticed during his lifetime, and have seldom been reprinted since. "It has always pleased me to exalt plants in the scale of organised beings," he wrote. He didn't care much anymore whether his books astonished the world and earned bagfuls of money. He was immune to ambition now and enamored, as ever, with the beautiful significance of tiny details and the big truth of interconnectedness.

42

The family home, Down House, was a quieter place in early 1876 than it had been for years. On February 12, Darwin turned sixty-seven. He and Emma weren't quite empty-nesters but, as events would unfold, they were as close to that circumstance as they'd ever get. William, the oldest son, was a banker down in Southampton, wise with his money, going bald like his father, and unmarried. Annie was dead and buried at Malvern. Little Charles and the infant daughter from the early 1840s, Mary Eleanor, were planted in the churchyard at Downe. Henrietta, the oldest surviving daughter, had married a slightly peculiar man named Litchfield and settled in London. After five years, the Litchfields were childless. George and Francis (usually called Frank) and by now even Horace, the youngest surviving son, had finished Cambridge and were variously groping, as their father had groped, toward serious interests and careers; only Frank, with a medical degree but no desire to practice, had returned to live near his parents. Leonard, who considered himself the dolt of the family, went to Woolwich Military Academy instead of Cambridge, then off on travels as a military engineer. That left just Bessy, the youngest daughter, unmarried at twenty-nine and destined to remain that way. Never educated outside the house, Bessy was "not good at practical things," according to one loving relative, and couldn't have managed her own life without help. Even Henrietta bullied her. She was so overlooked and inconsiderable, poor woman, that she scarcely gets mentioned in even the most thorough Darwin biographies. The record doesn't seem to tell Bessy's whereabouts in early 1876, but there's nowhere she could have been except at home.

The place didn't clatter with youth and energy, as it once had. There were no grandchildren to renew the cheerful chaos. Darwin worked on his plants and played backgammon with Emma in the evenings. The backgammon was an old tradition. Over the years he had won (as he bragged, self-mockingly) 2,795 games to her piddling 2,490. Frank, after hours, walked back down the lane to the little house where he lived with his wife, Amy. Even the faithful old butler, Parslow, was gone. He'd retired on a modest Darwin-paid pension.

Darwin himself, the elderly invalid, depended more than ever on Emma, the caregiver, doting spouse, and emotional pillar of the family. He might tease her about backgammon, but his devotion to this woman, which had started so tepidly back in 1838, had grown warm and fervent with passing time. She didn't share his intellectual interests; she didn't share his disdain for religion or his materialistic view of the world; she still worshiped a Christian God and worried for the soul of her husband; and he, for his part, loved her to the moon. He couldn't pretend to endorse her beliefs, or to accept his own deepest sorrows (such as the loss of Annie) and illnesses in a spirit of pious surrender, as Emma wanted him to do. But he venerated her goodness and was sensitive to her feelings. For forty years, he had saved—somewhere amid his portfolios and papers—that earnest letter she had written to him around the time of their marriage, after hearing his confession of wild and heretical ideas. "Don't think that it is not my affair," she'd insisted, "and that it does not much signify to me." It did signify, she wrote, arguing firmly but lovingly against his apostasy. Everything that concerned him concerned her also, and she would be "most unhappy if I thought we did not belong to each other for ever." Over all the years since, she had been left to nurture her hope of eternal togetherness, in an afterlife, without receiving any affirmation of that hope from him. Darwin could only sympathize, or avoid the subject; it wasn't in him to lie. But sometime along the way he had scribbled a note at the bottom of her letter. It was found there, among his other papers, after the end:

When I am dead, know that many
times, I have kissed and cryed over
this. C.D.

They were an old married couple, cousins and lovers and friends, hearing the echo of their own steps in a large cold house, as they moved inexorably toward death and separation.

And then, in the spring, came a bit of good news: Amy, Frank's wife, was pregnant. This provided some incentive for Darwin's next literary project. Faced with the prospect of a grandchild at last, he began to draft his private autobiography, hoping it "might possibly interest my children or their children." He wrote the first pages in late May 1876, during a visit at the country home of Emma's brother, and he continued back at Downe, spending an hour or so on the manuscript most afternoons through the summer. He searched his memory for signal facts and episodes, without consulting (for a change) any notebooks, portfolios of data, or diaries. My mother died when I was eight years old, Darwin wrote, and "it is odd that I can remember hardly anything about her" except her deathbed and one black velvet gown. But he recalled the time when, as a small boy, he stole apples from an orchard; he recalled his early collections of shells, birds' eggs, and minerals; he recalled, with indelible guilt after sixty years, having once been cruel to a puppy; and he recalled his undistinguished performance in boarding school, bad at Greek, bad at Latin, conscientious but never much engaged. He remembered his passion for bird-hunting, especially the moment of trembling excitement after he'd killed his first snipe. And he couldn't forget that caustic remark by his father, probably made when Charles was a teenager: "You care for nothing but shooting, dogs, and rat-catching, and you will be a disgrace to yourself and all your family." He remembered Robert Grant's guiding influence in Edinburgh, and then the rowdy gang of boozing cardplayers he fell in with at Cambridge. He remembered some loftier diversions during the Cambridge years, too—learning botany from Henslow, hearing the choir at King's College Chapel. "But no pursuit at Cambridge was followed with nearly so much eagerness or gave me so much pleasure," Darwin wrote, "as collecting beetles."

He hadn't dissected them. It was a fancy, not a way of doing science. He hadn't wondered about their geographical distribution or their morphological homologies—not in those days. He'd simply acquired them, identified them, and treasured them. Looking back from old age, he could still recollect particular species that had delighted him—such as *Panagaeus cruxmajor*, an orange-and-black ground beetle—and the rotting trees and dirt banks where he'd caught them. He told a story on himself that showed off his goofy zeal: Hunting beetles one day in some dead bark, he'd spotted a rare kind, then another, and grabbed one in each hand; then he'd seen "a third and new kind, which I could not bear to lose, so that I popped the one which I held in my right hand into my mouth. Alas it ejected some intensely acrid fluid, which burnt my tongue so that I was forced to spit the beetle out," allowing that specimen and the third to get away. A beetle-flavored pratfall.

Telling such stories was the easy, light side of autobiography. The harder part was asking himself: *What sort of man, with what strengths and weaknesses, what convictions, what doubts, have I been?* He did that, too.

None of this was intended for publication. After nine weeks of casual work, he tucked the manuscript away and returned to his plants. Several times over the next few years he pulled it back out and wrote more, inserting fresh recollections and afterthoughts, updating the section where he reflected on his published books. In finished form, with its artless candor and conversational tone, it offers a readable mix of personal narrative, portraits from memory, and philosophical self-examination. Its most revealing section (omitted from Darwin's own careless table of contents, as though he halfway intended to leave this part invisible) is titled "Religious Belief." He must have struggled with some tension between delicacy and bluntness, as he committed these thoughts to paper, because of the gap between his views and Emma's.

Before the idea of transmutation occurred to him, Darwin recollected, he'd been a very orthodox young man. Aboard the *Beagle* he was teased for his Bible-quoting piety. While conceiving his theory in the notebooks, during the late 1830s, he had thought much about religion. But years of studying the fixed laws of nature had eroded his credence in miracles, and then he "gradually came to disbelieve in Christianity as divine revelation." There was no smugness and no hastiness to his loss of faith; it happened almost against his will. "Thus disbelief crept over me at a very slow rate, but was at last complete." The change came so slowly, in fact, that he'd felt no anxiety. And now that it was done, he harbored no doubts. He added sternly:

> I can indeed hardly see how anyone ought to wish Christianity to be true; for if so the plain language of the text

seems to show that the men who do not believe, and this would include my Father, Brother, and almost all my best friends, will be everlastingly punished.

And this is a damnable doctrine.

To Emma, that passage would be especially disturbing. (She suppressed it from the first published version of the autobiography, five years after Darwin's death.) Darwin must have foreseen her reaction and felt badly about it, but in the privacy of the manuscript, once and for all, he was speaking his mind.

Beyond renouncing Christian dogma, he had given up any general belief in a personal God. What about the existence of evil, in a world supposedly run by a benevolent, omnipotent deity? That's so illogical, Darwin wrote, that "it revolts our understanding." What about the immortality of the human soul? That's a comforting notion we tend to embrace by instinct, he suggested, because the alternative is too dark to contemplate. What about the ultimate origin of life? The birth of the universe? Is there an ethereal First Cause, an abstract and impersonal Ultimate Being that set the world and its laws into motion? "I cannot pretend to throw the least light on such abstruse problems," Darwin admitted. "The mystery of the beginning of all things is insoluble by us; and I for one must be content to remain an Agnostic." His friend Huxley had invented the very word "agnostic," and Darwin was glad to have it. He felt that the alternative, "atheist," was too aggressive and confident.

The whole theme of this oddly titled section, "Religious Belief," is his lack of any such thing. His early religious beliefs had all been supplanted by guiding convictions of another kind. He looked back at William Paley's old argument from

design, as given in *Natural Theology* and once admiringly
absorbed by Darwin himself as a young Cambridge student.
The design argument fails, he declared flatly, "now that the law
of natural selection has been discovered." The beautiful hinge
of a bivalve shell, unlike the hinge of a door, doesn't imply the
existence of an intelligent designer. "There seems to be no
more design in the variability of organic beings and in the
action of natural selection," Darwin wrote, "than in the course
which the wind blows."

43

His first draft of the autobiography was finished and set aside
by early September 1876, when Frank's wife, Amy, went into
labor. She gave birth to a healthy son but then spiraled away
herself, probably from puerperal fever. After horrific convul-
sions and kidney failure, she died, with her husband and
father-in-law both there to watch her go. Darwin hadn't
absented himself this time from the difficult scene. Frank
buried Amy in Wales, among her father's people, and then
returned to vacate their little house, too grim with memories
and emptiness now. He and the baby boy, named Bernard,
moved back across the village into the big family home.

So the first Darwin grandchild grew through his infant and
toddler years in the close presence of a doting grandfather.
Unlike so many of Darwin and Emma's own children, Bernard
was fat and healthy as a baby, "a prize article" with a sort of
Buddha calm. "He has a pretty mouth and expression," Emma
thought, "and is particularly amused at his grandfather's face."
Darwin was reciprocally amused. He didn't turn Bernard's
growth and development into a research topic, as he had with

his own firstborn son; he simply enjoyed the little guy. Emma and he ordered renovations on the house so as better to accommodate Frank (with his own working life, as Darwin's assistant) and Bernard. The four of them went on a vacation together (with Etty and probably Bessy in tow also), traveling by private railway car to the Lake District. Bernard and Darwin developed pet names for each other—grandpa was "Baba" and Bernard was, for some nonsense reason, "Abbadubba." At the age of five, Abbadubba was welcome to amuse himself quietly on the floor of Darwin's study, drawing pictures, while Baba worked. They shared walks around the garden and stood hand-in-hand on the lawn during an outdoor concert. Within the next few years Bernard was destined to change quickly, of course, growing past the cute toddler stage into a lanky Etonian teenager, but Darwin wouldn't live to see that.

By late 1881, he was feeling heart pains. And now it wasn't some mysterious chronic ailment or hypochondria.

Unlike the butler Parslow, he never retired. Since he didn't hold a job or a position—instead, he was held *by* a vocation—there was nothing for him to quit. His work was the heartbeat of his life. He hated idleness even more than he hated exhaustion. He needed a project, always. His last major research effort involved earthworms and their role in making soil, which led to that small, quirky book mentioned earlier, *The Formation of Vegetable Mould, through the Action of Worms, with Observations on Their Habits*, published the year before he died. He had first addressed the subject in 1837, when he was fresh off the *Beagle*, and then returned to it forty years later, after the more urgent tasks of his scientific destiny had been fulfilled. He liked earthworms. They met his standards for a good research subject: humble creatures, nearly ubiquitous, and more important than they seem, producing tiny incremental effects with big cumulative consequence. He kept some in his study, potted like plants, and performed all manner of dotty experiments. There was no aspect of earthworm behavior that didn't interest him. "Worms do not possess any sense of hearing," he wrote in the book.

> They took not the least notice of the shrill notes from a metal whistle, which was repeatedly sounded near them; nor did they of the deepest and loudest tones of a bassoon. They were indifferent to shouts, if care was taken that the breath did not strike them. When placed on a table close to the keys of a piano, which was played as loudly as possible, they remained perfectly quiet.

Frank was the bassoonist. Emma played the piano. Little Bernard, according to one reliable source, commanded the whistle. Worm research, Darwin style, was an activity for the whole family. He established, with their help, that earthworms aren't musical.

The Formation of Vegetable Mould came out in October 1881, and, to the amazement of both Darwin and John Murray, all copies of the first printing and then the second were snatched up fast. Murray reprinted three more times, a thousand copies each time, before the end of the year. Readers evidently were ready for such a plainspoken, earthy volume from the formidable Mr. Darwin. Its brevity was another recommending feature. The book sold as though Darwin's homey but effective brand of science, combining extraordinarily keen and patient observation with kitchen-simple experimental methods, was going out of style. As a matter of fact, it was.

44

He died on the afternoon of April 19, 1882, of degenerative heart failure, at the age of seventy-three. Emma and Frank and Henrietta and Bessy were at his bedside. Bernard, in the nursery, knew only that Baba was somehow ill. The end wasn't tranquil; he suffered pain, nausea, and spasms that left him retching blood. His white beard was streamed and sticky with red. Between the attacks he caught his breath weakly. At one point he said, "I am not the least afraid to die," knowing people would wonder. At another moment, he whispered to Emma, "My love, my precious love." After several hours, he muttered, "If I could but die," and repeated the phrase like a plea, trying to let go. He dozed, he woke; they gave him a few spoonfuls of whiskey; he felt faint, and blacked out again. Then he was gone—gone in more senses than one. He left Downe in a horse-drawn hearse, headed for London.

The world had swooped in and claimed his body for history and posterity and the glory of British culture, et cetera, by way of a hastily assembled consensus among government officials and his scientific friends. The consensus decreed that Charles Darwin be buried in Westminster Abbey, like Lyell, like Newton and Chaucer, not in the village churchyard amid Erasmus and little Charles and Mary Eleanor and the good Kentish worms, as he might have preferred. If there was ever a funeral from which Darwin would have absented himself, given the chance to dodge out, it was his own. Too much fuss. Bad for the stomach. Queen Victoria skipped it herself, and so did Prime Minister Gladstone; but Parslow was there. The pallbearers included Hooker and Huxley and Wallace.

Before all this occurred, though, Darwin had finished one more small piece of work. The worms were his final book but not his final publication. In the weeks before his death, he had turned back to another early interest: means of dispersal, by which animal and plant species colonize new places.

It's the first premise of modern biogeography, crucial to his evolutionary theory, that patterns of species distribution reflect natural dispersal from points of evolutionary origin, not whimsical geographic assignments from the hand of God. Darwin had investigated such means of dispersal, back in the 1850s, with his experiments involving saltwater immersion and other forms of simulated environmental travail. He had rafted asparagus across tiny seas. He had shoved seeds into the bellies of dead fish, fed the fish to pelicans, then collected the pelican poop and extracted the seeds to see whether they retained their capacity to germinate. He had dangled duck feet in an aquarium full of freshwater snails, inviting the adventuresome to take hold. And now he found fascination in a similar piece of data, derived not from experiment but from accidental observation. Somewhere near Northampton, in a stream or a pond, a small freshwater clam had clamped itself onto the leg of a water beetle.

One beetle, dragging a miniature mollusk. Extracted from its scientific context—from the questions of dispersal and biogeography and evolution versus special creation—it seems utterly insignificant. Maybe it seems insignificant *within* its context. But it didn't to Darwin. He described the beetle-clam connection in a short note to *Nature*, which appeared on April 6, 1882. Titled "On the Dispersal of Freshwater Bivalves," it was his last published work. The point was simple but substantive: Here's evidence of how a fertile clam might travel by air (since water beetles fly as well as swim) from one pond to another and establish a transplanted population in a new place. Dispersal, biogeography. Colonization, and then a fresh phase of evolution.

He had gotten his raw data, as he'd always gotten so much, through the mail. A young man named W. D. Crick had written from Northampton, bringing this little coleopterous discovery to Darwin's attention. Mr. Crick (whose eventual grandson, Francis Crick, would play his own sizable role in the history of biology as co-discoverer of the structure of DNA) was a rising shoe-factory owner with a fondness for nature. The beetle was *Dytiscus marginalis*, a large diving predator. What species of clam? W. D. Crick didn't know. When Darwin wrote back, asking further questions, Mr. Crick obliged him by sending the beetle itself, and the clam, by return post. But the two creatures were no longer attached. Being out of water was stressful for both of them, and the clam (the "shell," as Darwin and Crick called it) had dropped away and slammed itself tight.

"I have placed the shell in fresh-water, to see if the valve will open," Darwin informed Crick. He wanted to know whether, like the briny asparagus seeds, it was still viable after its travels.

There were matters of science at issue, and there were matters of life, common decency, mercy. "As the wretched beetle was still feebly alive," he told Crick, "I have put it in a bottle with chopped laurel leaves, that it may die an easy and quicker death." Any naturalist of his day knew that laurel leaves, when chopped, release prussic acid, containing hydrogen cyanide. Darwin didn't want his last beetle to suffer. He was a gentle man, quite aware that he'd already caused discomfort enough.

Source Notes

For the sake of clarity and simplicity, Darwin's writings are cited by abbreviated title. *Correspondence* indicates the series edited by Burkhardt, et al. A citation of "*Correspondence* 3: 43," for instance, directs you to page 43 of the third volume, covering the years 1844–46. *CD's Notebooks* is the compilation by Barrett, et al. (1987). *Autobiography* refers to the Barlow (1969) edition. For citations from *The Origin of Species*, first edition, I've used a facsimile reprint of that edition (Harvard University Press, 1964, with an introduction by Ernst Mayr), cited as *The Origin*. Other sources are cited by author and year, or as descriptively necessary.

In quoting from Darwin's letters and private notes, I have in a very few instances made small corrections to spelling, punctuation, or grammar.

Home and Dry

11   ***the face on the dollar bill***: as it is, currently, on the Bank of England's £10 note.

13   ***"People often talk"***: *CD's Notebooks*, 222–23.

14   ***"God created humans pretty much"***: Gallup News Service, November 19, 2004, and earlier poll reports from Gallup.

The Fabric Falls

21   *"a disgrace to yourself"*: *Autobiography*, 28.

21   *"Why, the shape of his head"*: Ibid., 79.

22   *"writing is most tedious & difficult"*: *Correspondence* 2: 39.

25   *"If there is the slightest foundation"*: *Beagle Diary*, 357, note 1, quoting from CD's ornithological notes.

27   *"These facts origin (especially latter)"*: *Correspondence* 2: 431.

27   *called his notebooks on the "transmutation" of species*: Ibid., 2: 431.

27   *he wrote "Zoonomia," in genuflection*: *CD's Notebooks*, 170.

28   *"all warm-blooded animals have arisen"*: Ibid., note 1.5.

28   *"Why is life short"*: Ibid., 171.

28   *One result: "adaptation"*: Ibid.

29   *"who will dare say what result"*: Ibid., 172.

29   *"According to this view animals"*: Ibid.

29   *"Each species changes"*: Ibid., 175.

29   *he drew a rough diagram of a lineage*: Ibid., 180.

29   *"Heaven knows whether this agrees"*: Ibid., 181.

30   *"Apteryx," he wrote, "a good instance"*: Ibid., 211.

32   **"Species have a real existence in nature"**: Quoted in Hull (1983), 68.

33   *"Seldom in the history of ideas"*: Ibid., 15.

36   *"The changes in species must be very slow"*: *CD's Notebooks*, 242.

36   *"all the change that has been accumulated"*: Ibid., 248.

36   *comments in the notebook about "my theory"*: Ibid.

36   *"Study the wars of organic being"*: Ibid., 262.

36   *"a most laborious, & painful effort"*: Ibid., 263.

37   *the "whole fabric totters & falls"*: Ibid.

37   *"But Man—wonderful Man"*: Ibid.

37   *concluding firmly that, no, "he is no exception"*: Ibid.

38   *"I have not been very well of late"*: *Correspondence* 2: 47.

38   *"anything which flurries me completely knocks me up"*: Ibid., 2: 51–52.

38   *he repeated to his old friend W. D. Fox*: Ibid., 2: 91.

39   *"If not marry," he wrote*: Ibid., 2: 443.

40   *"If marry," he wrote*: Ibid.

41   *this time "Marry" headed the longer column*: Ibid., 2: 444.

41   *"It being proved necessary"*: Ibid.

41   *"Mine is a bold theory"*: *CD's Notebooks*, 340.

42   *"There must be some law"*: Ibid., 347.

42   *he "thought much upon religion"*: *Correspondence* 2: 432.

43   *"their crowding and interfering with each other's"*: Wrigley and
        Souden, eds. (1986), 7.

44   *prevented what Malthus called "checks"*: Ibid., 14.

44   *"moral restraint, vice, and misery"*: Ibid., 16.

44   *"the warring of the species as inference"*: *CD's Notebooks*, 375.

45   *"One may say there is a force"*: Ibid., 375–76.

45   *a burst of comments on the "differences"*: Ibid., 391.

45   *increasingly confident references to "my theory"*: Ibid., 397–98.

46   *labeled "N," which was devoted to "metaphysical enquiries"*:
        Ibid., 561.

46   *whether "love of the deity" might result*: Ibid., 291.

46   *"Having proved mens & brutes bodies"*: Ibid., 409.

47   *He wanted to illuminate those "laws of life"*: Ibid., 411.

48   *Emma felt "bewildered" rather than giddy*: Quoted in Browne
        (1996), 391.

49   *she called it "a painful void between us"*: *Correspondence* 2: 123.

49   *his "most sincere love & hearty gratitude"*: Ibid., 2: 114.

50   *"Three principles, will account for all"*: *CD's Notebooks*, 412–13.

50   *"Wasted entirely the last week of November"*: *Correspondence* 2:
        432.

50   *"stupid & comfortable"*: Ibid., 2: 156.

The Kiwi's Egg

53   *a multi-part paper titled "On the Anatomy of the Apteryx"*:
        Owen (1838).

53   *referred to it in his notebook "D"*: *CD's Notebooks*, 340–41.

54   *"Sometimes the egg-bearing female"*: Grzelewski (2000), 83.

57   *"while you are acting conscientiously"*: *Autobiography*, 235–37.

58 *This "sketch," as he called it*: *Correspondence* 2: 435.

59 *what he called here "the natural means of selection"*: F. Darwin, ed. (1909), 4.

59 *would be to "alter forms"*: Ibid., 9.

60 *with their "deceptive appearance"*: Ibid., 49.

60 *"not from one law of gravity"*: Ibid.

60 *"death, famine, rapine, and the concealed war"*: Ibid., 52.

60 *"There is a simple grandeur"*: Ibid.

61 *"impious doctrines" such as atheism and socialism*: Desmond and Moore (1991), 297, quoting *The Times* in August 1842.

63 *to his "dear old Titty" or his "dear Mammy"*: *Correspondence* 2: 398; Ibid., 4: 146.

64 *"classification consists in grouping beings"*: Ibid., 2: 376.

65 *"a most strange fact"*: Ibid., 3: 421.

66 *asking Hooker's help with "one little fact"*: Ibid., 3: 2.

66 *"I have been now ever since my return"*: Ibid.

66 *"Heaven forfend me from Lamarck"*: Ibid.

68 *with its casual suggestion about "one living filament"*: *CD's Notebooks*, 170, note 1.5.

70 *Lamarck thought, by "the supreme author of all things"*: Quoted in Mayr (1982), 353.

70 *certain "subtle fluids" somehow*: Burkhardt, ed. (1977), 155–56; Bowler (1989), 84.

71 *Lamarck's sentiment intérieur*: Burkhardt, ed. (1977), 169.

71 *a sort of "feeling of existence"*: Packard (1901), 325.

72 *one particular organism, the "sea mat"*: Desmond and Moore (1991), 37; *Autobiography*, 50.

73 *"I listened in silent astonishment"*: *Autobiography*, 49.

74 *it "has met with some degree of favour"*: Lyell (1989), 2: 18.

75 *"& also a gradual change of species"*: *Correspondence* 3: 7.

76 *intended for reading "in case of my sudden death"*: Ibid., 3: 43.

77 *"it will be a considerable step in science"*: Ibid.

78 *somehow leads to the "creation or production"*: Ibid., 3: 61.

80 *"I have continued steadily reading"*: Ibid., 3: 67.

80 *"The general conclusion at which I have slowly"*: Ibid.

81   *"It has pleased Providence to arrange"*: Chambers (1994), 234.

82   *he seemed to be . . . a "funny fellow"*: *Correspondence* 3: 103.

82   *he'd been "somewhat less amused"*: Ibid., 3: 108.

82   *Darwin realized that "Mr. Vestiges"*: Ibid., 3: 253.

Point of Attachment

84   *"Darwin's Delay"*: Gould (1977), 21.

88   *"Seeing this gradation and diversity of structure"*: *Journal of Researches* (1845), reprinted as *The Voyage of the Beagle* (New York: Modern Library, 2001), 339.

88   *about how "the creative power"*: *Journal of Researches* (1839), facsimile reprint (New York: Hafner, 1952), 474. See also pp. 461–62 for Darwin's cursory 1839 comments on diversity of beak shape among the finches.

88   *marveling at "the amount of creative force"*: *Journal of Researches* (1845, 2001), 355.

88   *he admitted feeling "astonished"*: Ibid., 337.

88   *"Hence, both in space and time"*: Ibid.

89   *"the replacement of extinct species by others"*: *CD's Notebooks*, 413, note 59.2.

89   *made him a landlord, "a Lincolnshire squire"*: *Correspondence* 3: 181.

91   *"My life goes on like Clockwork"*: Ibid., 3: 345.

92   *"How much time lost by illness!"*: Ibid., 3: 397.

93   *"Where does your father do his barnacles?"*: Browne (1996), 473.

94   *"I am not inclined to take much for granted"*: *Correspondence* 3: 250.

94   *"How painfully (to me) true"*: Ibid., 3: 253.

94   *mentioning it fondly as "Mr. Arthrobalanus"*: Ibid., 3: 356.

100   *Geoffroy discerned "unity of plan"*: Desmond (1992), 47; Mayr (1982), 362.

102   *"lies in our ignorance of what"*: *Correspondence* 2: 375.

102   *that classification should reflect "consanguinity"*: Ibid., 2: 376.

103   *that Gray "intended to anticipate" Darwin's work*: Ibid., 4: 162.

104   *or as Charles sometimes called him, "the Governor"*: See, e.g., ibid., 2: 86, 118.

105   *"the kindest man I ever knew"*: *Autobiography*, 28.

106   *"unusually unwell, with swimming of head"*: *Correspondence* 4: 384.

107   *"My own ever dear Mammy"*: Ibid., 4: 183.

107   *rationalized that it was "only a ceremony"*: Ibid., 5: 9.

107   *rudimentary or "abortive" structures*: Ibid., 3: 365.

108   *"I have been struck"*: Ibid., 4: 344.

109   *"Systematic work wd be easy were it not"*: Ibid.

109   *"My cirripedial task is an eternal one"*: Ibid., 4: 319.

109   *"a small, poor first fruit"*: Ibid., 4: 344.

110   *"of which creatures I am wonderfully tired"*: Ibid., 5: 100.

110   *"a shout of paeans for the Barnacles"*: Ibid., 4: 406.

110   *The Royal Medal was "quite a nugget"*: Ibid., 6: 55.

110   *"Health very bad with much sickness"*: Ibid., 4: 385.

111   *caused "nervous dyspepsia"*: Ibid., 4: 225, note 1.

111   *"sugar, butter, spices tea bacon"*: Ibid., 4: 224.

111   *"I am turned into a mere walking"*: Ibid., 4: 234.

112   *"that it induces in most people"*: Ibid., 4: 235–36.

113   *he'd gotten to like his own "aquatic life"*: Ibid., 4: 344.

113   *"Her whole mind was pure & transparent"*: Ibid., 5: 541.

114   *worried that she had inherited his "wretched digestion"*: Ibid., 5: 9.

114   *A modern scholar named Randal Keynes*: See Keynes (2002), 241–45.

115   *"her face lighted up"*: *Correspondence* 5: 13.

115   *she vomited badly, and "from hour to hour"*: Ibid., 5: 14.

115   *Annie's "hard, sharp pinched features"*: Ibid., 5: 16.

115   *"I quite thank you"*: Ibid., 5: 19.

116   *The governess immediately had "one of her attacks"*: Keynes (2002), 216.

116   *"I am in bed not very well with my stomach"*: *Correspondence* 5: 24.

116   *"She went to her final sleep most tranquilly"*: Ibid., 5: 24.

117   *"We shall be much less miserable together"*: Ibid., 5: 25.

117   *"Sometime," he added, he would like to know*: Ibid., 5: 28.

118   *"I never gave up Christianity until I was"*: Desmond and Moore (1991), 658.

118   *Christianity was "not supported by evidence"*: Browne (2002), 484.

118   *"disbelief crept over me at a very slow rate"*: *Autobiography*, 87.

119   *"Everything in nature," he concluded coldly*: Ibid.

119   Later, *"with many fluctuations"*: Ibid., 93.

119   *he simply called himself "muddled"*: *Correspondence* 8: 275.

120   *attacked by Leibniz as "subversive" of natural religion*: Ibid., 8: 106, Darwin quoting Brewster's life of Newton.

120   **Gravity was "an occult quality"**: Ibid.

120   *"I cannot believe that there is a bit more"*: Ibid., 8: 258.

120   *"I cannot see, as plainly as others do"*: Ibid., 8: 224.

120   *"An innocent & good man stands"*: Ibid., 8: 275.

121   *"Blessings on her"*: Ibid., 5: 542.

A Duck for Mr. Darwin

122   *"the theory of the progressive development"*: Wallace (1905) 1: 254.

124   *Wallace found his book "very philosophical"*: Ibid.

126   *"shining out like a mass of brilliant flame"*: Wallace (1889), 152.

129   *"good resolutions soon fade"*: Wallace (1905), 1: 309.

132   *What he termed "closely allied species"*: Wallace (1852), reprinted in Camerini, ed. (2002), 101.

135   *"only the announcement of the theory"*: Marchant (1975), 54.

135   *He also wrote about the "creation" of new species*: Wallace (1969), 8–10.

136   *He used the term "antitype"*: Ibid., 6.

136   *"are all explained and illustrated by it"*: Ibid., 18.

136   *"Such phenomena as are exhibited by the Galapagos Islands"*: Ibid., 8.

137   **"Every species has come into existence coincident"**: Ibid., 6.

138   *"began sorting notes for Species Theory"*: *Correspondence* 5: 537.
139   *what he called "accidental means" of plant species dispersal*: *The Origin of Species*, facsimile 1st edn. (1964; cited hereafter as *The Origin*), 358.
140   *putrid and stinky "in a quite extraordinary degree"*: *Correspondence* 5: 332.
140   *"a wonderful quantity of mucus"*: Ibid., 5: 305.
142   *"I have puppies of Bull-dogs & Greyhound in salt"*: Ibid., 5: 326.
143   *"I am getting on splendidly"*: Ibid., 6: 45.
143   *"Skins," it began: "Any domestic breed or race"*: Ibid., 5: 510.
144   *Halfway down the list appeared an inconspicuous*: Ibid.
144   *"What think you of Wallace's paper"*: Ibid., 5: 519.
145   *but offered "nothing very new"*: Ibid., 5: 522.
145   *"It seems all creation with him"*: Ibid.
146   *"The domestic duck var. is for Mr. Darwin"*: Ibid., 6: 290.
147   *"By your letter, & even still more"*: Ibid., 6: 387.
149   *"I wish you would publish some small fragment"*: Ibid., 6: 89.
149   *"I rather hate the idea of writing"*: Ibid., 6: 100.
149   *"I had a good talk with Lyell"*: Ibid., 6: 106.
150   *though not necessarily against a "preliminary essay"*: Ibid., 6: 109.
151   *"I am working very steadily at my big Book"*: Ibid., 6: 265.
152   *"This summer will make the 20th year (!)"*: Ibid., 6: 387.

His Abominable Volume

153   *"On the Tendency of Varieties to Depart Indefinitely"*: Wallace (1858), reprinted as part of the full Darwin-Wallace presentation to the Linnean Society, in Weigel, ed. (1975). The Wallace paper alone is also reprinted in Wallace (1969).
154   *posited "a general principle in nature"*: Weigel, ed. (1975), 28.
154   *"The life of wild animals"*: Ibid.
154   *"the weakest & least perfectly organized"*: Ibid., 32.
154   *"variations from the typical form of a species"*: Ibid., 34.
155   *Those creatures "best adapted" as a result*: Ibid., 32.

155   *The result would be "continued divergence"*: Ibid., 36.

155   *"successive variations departing further and further"*: Ibid., 28.

155   *"all the phenomena presented by organized beings"*: Ibid., 42.

155   *"Ternate, February 1858"*: Ibid.

156   *"And the answer," as he recollected long afterward*: Wallace (1905), 1: 362.

156   *"The more I thought over it"*: Ibid.

158   *"Your words have come true with a vengeance"*: *Correspondence* 7: 107.

159   *"without its full share of intelligence"*: Litchfield (1915), 2: 162.

159   *describes baby Charles as "severely retarded"*: Desmond and Moore (1991), 467.

159   *"may have been slightly retarded"*: Browne (2002), 37, 503.

159   *"It was the most blessed relief"*: *Correspondence* 7: 121.

160   *"after their first sorrow, they could only feel"*: Litchfield (1915), 2: 162.

160   *"nice little bubbling noises"*: *Correspondence* 7: 521.

160   *"I shd. be extremely glad now to publish"*: Ibid., 7: 117.

161   *"This is a trumpery letter"*: Ibid., 7: 118.

161   **"I daresay all is too late"**: Ibid., 7: 122.

163   *though Darwin did allude to "the origin of species"*: Abstract of Darwin's letter to Asa Gray, printed in Darwin and Wallace (1858), reprinted in Weigel, ed. (1975), 22.

164   *that Darwin and Wallace's paper "would not be worthy of notice"*: *Correspondence* 7: 292.

164   *savored it as "a taste of the future"*: Ibid., 7: 291.

164   *"I shall never forget the impression"*: Quoted in Browne (2002), 49.

164   *references to the "ingenious and original reasonings"*: Quoted in ibid., 50.

164   *"any of those striking discoveries which at once"*: Quoted in ibid., 42.

165   *"I am quite prostrated & can do nothing"*: *Correspondence* 7: 121.

165   *he seized the idea of writing a sleek "abstract"*: Ibid., 7: 130.

166   *"When we look to the individuals"*: *The Origin*, 7.

167   suffering *"the old severe vomiting"*: *Correspondence* 7: 247.

167   *"My abstract is the cause"*: Ibid., 7: 247.

168   *"a comparatively free man"*: Ibid.

168   *"I had absolutely nothing whatever to do"*: Ibid., 7: 240.

168   *"written in 1839 now just 20 years ago!"*: Ibid., 7: 241.

168   *"perfect, quite clear & most courteous"*: Ibid., 7: 129.

169   if Mr. Darwin's *"excess of generosity" had resulted*: Ibid., 7: 166.

170   *"strictly just to both parties"*: Ibid.

170   *"who thought so highly of it that they immediately"*: Marchant (1975), 57.

170   *"This assures me the acquaintance and assistance"*: Ibid.

171   *"This insures me the acquaintance of these eminent men"*: Wallace (1905), 1: 365.

171   *"incredibly bad, & most difficult"*: *Correspondence* 7: 303.

171   *His proposed title was deadly dull*: Ibid., 7: plate facing p. 283.

172   *"So much for my abominable volume"*: Ibid., 7: 336.

172   *"no possible means of drawing [a] line"*: Ibid., 7: 349.

173   *their skills at "the American game"*: Ibid.

173   *"You cannot think how refreshing it is"*: Ibid., 7: 350.

174   *"My dear Sir, I have received"*: Ibid., 7: 365.

174   *"Your father says he shall never think"*: Ibid., 7: 395, note 1.

175   *"If a monkey has become a man"*: Peckham (1959), 16.

179   *"the Survival of the Fittest"*: Ibid., 22, 164.

180   *a whole new chapter, "Miscellaneous Objections"*: Ibid., 23, 226.

180   *struck the word "On" from his title*: Ibid., 23, 71.

181   *"When on board H.M.S. 'Beagle'"*: *The Origin*, 1.

182   *"innumerable species inhabiting this world"*: Ibid., 3.

182   *"perfection of structure and coadaptation"*: Ibid.

182   *that the mechanism he calls "natural selection"*: Ibid., 5. Darwin puts the phrase in initial capitals, "Natural Selection," at this first mention, but later (e.g., p. 81) he leaves it lowercase.

182   *"have been, and are being, evolved"*: Ibid., 490.

182   *talks in these opening pages about "modification and coadaptation"*: Ibid., 4.

182   *"descent with modification"*: Ibid., 331.

182   *or his "theory of descent"*: Ibid., 426.

182   *in his introduction is the "struggle for existence"*: Ibid., 4. Again, capitalized as "Struggle for Existence" on first use, but not always later, e.g., p. 5.

183   *"As many more individuals"*: Ibid., 5.

184   *"No one supposes that all the individuals"*: Ibid., 45.

185   *"much struck how entirely vague and arbitrary"*: Ibid., 48.

186   *"Darwin revolutionized our study of nature"*: Lewontin (2000), 67.

186   *"all nature is at war"*: The allusion to de Candolle is at *The Origin*, p. 62; the quote is from Darwin's 1844 essay, in F. Darwin, ed. (1909), 87.

187   *"The face of Nature may be compared to"*: *The Origin*, 67.

187   *"what may not nature effect"*: Ibid., 83.

188   *"In such case, every slight modification"*: Ibid., 82.

189   *"The more diversified the descendants"*: Ibid., 112.

191   *"every species has come into existence coincident"*: Ibid., 355, quoted from Wallace (1855).

191   *"We see in these facts some deep organic bond"*: *The Origin*, 350.

192   *"On the theory of descent with modification"*: Ibid., 340.

192   *"the animal in its less modified state"*: Ibid., 449.

192   *the "very soul" of natural history*: Ibid., 434.

193   *And the "same great law" of homologous parts*: Ibid.

193   *"that it has so pleased the Creator"*: Ibid.

194   *but somehow "natural" and objective*: Ibid., 413.

194   *simply "reveals the plan of the Creator"*: Ibid.

194   *"all true classification is genealogical"*: Ibid., 420.

194   *"the hidden bond which naturalists"*: Ibid.

195   *he also calls them "atrophied" or "aborted"*: Ibid., 450.

195   *the entire book is essentially "one long argument"*: Ibid., 459.

196   *"a considerable revolution in natural history"*: Ibid., 484.

196   *"In the distant future"*: Ibid., 488.

196   *"Light will be thrown on the origin of man"*: Ibid.

196   *"To my mind it accords better"*: Ibid.

197   *"There is grandeur," he says finally*: Ibid., 490.

198 *"This Abstract, which I now publish"*: Ibid., 2.

199 *"If I had space," he claims later*: Ibid., 31.

199 *" . . . but I have not space here"*: Ibid., 89.

199 *"I could show by a long catalogue of facts"*: Ibid., 45.

199 *"I shall reserve for my future work"*: Ibid., 53.

199 *"if I had space, I could show that they are conformable"*: Ibid., 230.

200 *"This view may be true, and yet it may never be capable"*: Ibid., 338.

200 *"the assertion is quite incapable of proof"*: Ibid., 468.

200 *"I think it highly probable that" and "I am convinced that"*: Ibid., 18, 43.

201 *The adjective "random" appears nowhere*: Attested by Barrett, et al. (1981), 606.

201 *to say that variations are "due to chance"*: The Origin, 131.

202 *sounds concrete and sensible: the "effects of use and disuse"*: Ibid., section heading, 134.

202 *"I think there can be little doubt that use"*: Ibid., 134.

202 *"I believe that the nearly wingless condition"*: Ibid.

203 *"The real triumph of Darwin's book"*: Peckham (1959), 25.

204 *"If I lived twenty more years"*: F. Darwin, ed. (1903), 2: 379.

The Fittest Idea

207 *variations occur in response to "conditions of life"*: The Origin, 131.

207 *"Our ignorance of the laws of variation"*: Ibid., 167.

208 *variations occur "in no determinate way"*: F. Darwin, ed. (1909), 85.

208 *he had described them as "accidents"*: CD's Notebooks, 633.

208 *to say they are "due to chance"*: The Origin, 131.

209 *"a dish of rank materialism cleverly cooked"*: Quoted in Hull (1973), 169.

209 *suggesting that "man might be a transmuted ape"*: Quoted in ibid., 177.

209   *positing instead some "internal innate force"*: Quoted in ibid., 409.

211   *"The 'Doctrine of Uniformity' in Geology Briefly Refuted"*: Proceedings of the Royal Society of Edinburgh, vol. 5, 1866, cited in Gould (2002), 492–93.

211   *far less than the "incomprehensibly vast"*: The Origin, 282.

212   *"sufficient to disprove the doctrine that transmutation"*: Quoted in Gould (2002), 497.

212   *about the "odious spectre" of Thomson*: Marchant (1975), 220.

212   *cut "incomprehensibly vast" to merely "vast"*: Peckham (1959), 478.

212   *"we have no means of determining how long"*: Ibid., 486.

213   *"Fleeming Jenkin has given me much trouble"*: F. Darwin, ed. (1903), 2: 379.

213   *"the tendency to revert to parent forms"*: CD's Notebooks, 248.

214   *"elaborated by Mr. Darwin in his celebrated"*: Wallace (1962), 102.

215   *"moral and higher intellectual nature of man"*: From "Sir Charles Lyell on Geological Climates and the Origin of Species," The Quarterly Review, 1869, vol. 126, excerpted in Smith, ed. (1991), 31.

215   *"that an Overruling Intelligence has watched"*: Ibid., 33–34.

215   *"I shall be intensely curious to read"*: F. Darwin, ed. (1903), 39.

216   *Darwin scratched "No!!!"*: Browne (2002), 318.

218   *"Lamarckism in a modern form"*: Packard (1980), 393, quoting himself from an earlier paper.

218   *"nearer the truth than Darwinism proper"*: Ibid.

218   *became known as "the law of acceleration"*: Bowler (1992), 128; Gould (2002), 367.

220   *a book about the "self-adaptation" of plants*: Quoted in Bowler (1992), 86.

221   *a claim that internal "laws of growth" dictate*: Quoted in ibid., 90.

221   *"that the actual course of orthogenetic evolution"*: Ibid., 151.

222   ***"must advance by the shortest and slowest steps"***: *The Origin*, 194.

223   ***"I know that you are studying hybrids"***: Quoted in Gould (2002), 419.

225   ***whereas others are* recessive *(Mendel's terminology)***: Mendel (1965), 8.

227   ***He argued that the* germ plasm**: Mayr (1982), 700.

231   ***"based on logic and on interpretation of many kinds"***: Futuyma (1998), 11.

231   ***"fully vindicate his hypothesis"***: Ibid., 12.

233   ***"about 30,000 genes, with 99%"***: *Nature*, vol. 420, December 5, 2002, 509.

233   ***The resemblance between our 30,000 human genes***: Dr. Futuyma was rounding off, based on the best gene counts available at that time. Since my conversation with him, on January 21, 2004, further work in genomics has led to a slight revision: Scientists now believe there are only about 20,000 to 25,000 human genes, rather than 26,000 to 31,000, as thought when the hurried results of the Human Genome Project were first announced. The mouse genome also continues to be studied; the overwhelming degree of mouse-human similarity hasn't been challenged; and the downward revision in gene number doesn't affect the validity of Futuyma's point.

233   ***"Comparative genome analysis is perhaps"***: *Nature*, vol. 420, December 5, 2002, 557.

The Last Beetle

237   ***with salutations from a "sincere admirer"***: Quoted in Desmond and Moore (1991), 601.

239   ***"interesting letter" about hairiness on the ears***: F. Darwin, ed. (1903), 2: 53–54.

240   ***"It has always pleased me to exalt"***: *Autobiography*, 135.

241   ***Bessy was "not good at practical things"***: Raverat (1952), 146.

242   *"Don't think that it is not my affair"*: Reprinted in *Autobiography*, 237.

243   *"When I am dead, know that many"*: Ibid.

243   *"might possibly interest my children"*: *Autobiography*, 21.

243   *"it is odd that I can remember hardly anything"*: Ibid., 22.

244   *"You care for nothing but shooting"*: Ibid., 28.

244   *"But no pursuit at Cambridge was followed"*: Ibid., 62.

244   *"a third and new kind, which I could not bear to lose"*: Ibid.

245   *is titled "Religious Belief"*: Ibid., 85, but absent from the table of contents, 19.

245   *"gradually came to disbelieve in Christianity"*: Ibid., 86.

245   *"Thus disbelief crept over me"*: Ibid., 87.

245   *"I can indeed hardly see how anyone"*: Ibid.

246   *"it revolts our understanding"*: Ibid., 90.

246   *"I cannot pretend to throw the least light"*: Ibid., 94.

246   *the alternative, "atheist," was too aggressive*: Desmond and Moore (1991), 657.

247   *"now that the law of natural selection"*: *Autobiography*, 87.

247   *"a prize article" with a sort of Buddha calm*: Quoted in Browne (2002), 438.

247   *"He has a pretty mouth and expression"*: Quoted in ibid., 438.

248   *grandpa was "Baba" and Bernard*: Ibid., 491.

249   *"Worms do not possess any sense of hearing"*: CD, *The Formation of Vegetable Mould*, 29.

250   *"I am not the least afraid to die"*: Desmond and Moore (1991), 662.

250   *"My love, my precious love"*: Ibid., 661.

250   *"If I could but die," and repeated the phrase*: Ibid., 662.

252   *"I have placed the shell in fresh-water"*: F. Darwin, ed. (1903), 2: 29.

252   *"As the wretched beetle was still"*: Ibid. 2: 29.

Bibliography

Charles Darwin was a prolific writer. His lifetime output includes the published books and articles, the private notebooks, and a formidable abundance of personal letters. The best evidence for who he was, and for what he thought, is what he wrote. Close attention to the text of the first edition of *The Origin of Species*, for instance, and to the changes he made in later editions, is the right substitute and remedy for blurred, secondary notions of what is or isn't "Darwinian." Darwin's letters are very illuminating, especially as edited and annotated by Frederick Burkhardt and a team of other scholars in the ongoing multi-volume series from Cambridge University Press. The transmutation notebooks, transcribed and edited by Paul H. Barrett and some colleagues, reveal much about how Darwin pieced his ideas together. His *Autobiography,* restored to completeness (after earlier suppression of some passages, in deference to Emma Darwin) and edited by his granddaughter, Nora Barlow, is also a telling document. These have been my primary sources. Confession: I haven't read every word that Darwin put into print. My selective list of his books, below, records only those that have been most influential scientifically and that were crucial or useful to me.

The secondary literature on Darwin and his work (sometimes called the "Darwin Industry" of scholarship and commentary) is huge and still grow-

ing. Every week, it seems, someone publishes another erudite research article or argumentative book with his name in the title. The list below is,
again, partial and personal. Think of it as merely a small sample, purpose-
ful but somewhat subjective on my part, of what's out there on the subject
of Charles Darwin.

In case you find yourself hungry for further reading—of Darwin's own
writings, or about him—I'll offer a few suggestions. For a start that takes
you straight to the core of the man and his work, read *The Origin of Species*,
preferably in a reprint of the first edition. *The Voyage of the Beagle* (as his
Journal of Researches has been titled in later editions) is also a wonderful
book in its way—not so important as *The Origin of Species* (how many
books are?) but more relaxed and amusing, a travel narrative filled with
acutely observed natural history and told in the voice of a likable, unpretentious young Englishman. *The Autobiography of Charles Darwin*, first
published in 1887 as part of *The Life and Letters of Charles Darwin*, and
later available in the Nora Barlow edition, is warm and personable (mainly
because he wrote it for his children, not for distant readers such as us),
graced by his gentle spirit and his extraordinary honesty. For those who
enjoy reading letters, you don't need to plunge into the oceanic thorough-
ness of that series from Cambridge; Frederick Burkhardt has also edited a
nice little book, *Charles Darwin's Letters: A Selection*. Better still, find a
reprint of Francis Darwin's original 1887 compilation, *The Life and Letters
of Charles Darwin*, in two modest-sized volumes. Volume I of that set gives
you, besides letters and the autobiography, a long chapter of reminiscences
by Francis about his father's character, working habits, and everyday life.

Among the many Darwin biographies, the two most impressive are also
the two most appealing: Janet Browne's and the one by Adrian Desmond
and James Moore. Both works are enriched by formidable scholarship and
enlivened by sharp insights and good writing. Desmond and Moore are
especially strong on the political context surrounding Darwin and his
thought. Browne's two-volume life is particularly good on the social milieu
in which Darwin lived, and on the high-minded ruthlessness with which he
presumed upon the females of his own family, as well as friends and min-
ions around the world, to indulge his disposition and serve his intellectual
demands. The biographical portrait by Darwin's great-great-grandson,

Randal Keynes, published in the United States as *Darwin, His Daughter, and Human Evolution* (originally in the UK as *Annie's Box*) is also a valuable book, poignant but levelheaded, and informed by insider material.

For Darwin's own works, what I cite below are the original bibliographical data—details of the first editions—so you can see at a glance how his publishing career unfolded. In my Source Notes, on the other hand, I place quotations from Darwin within the editions that were available to me. Sorry for this inconsistency, but on the whole I think it keeps things more clear. In the list of secondary material, I cite the editions that came into my hands. Some of those books and articles represent milestones in the development of evolutionary theory, published long ago and since reprinted. In such cases (for instance, Mendel's paper), where the year of first publication helps clarify the historical significance, I supply the original year in parentheses.

1. Works published by Charles Darwin

1839. *Journal of Researches into the Geology and Natural History of the Various Countries Visited by H.M.S. Beagle, Under the Command of Captain FitzRoy, R.N. from 1832 to 1836.* London: Henry Colburn.

1839. "Observations on the Parallel Roads of Glen Roy," *Philosophical Transactions of the Royal Society of London*, 39–81.

1842. *The Structure and Distribution of Coral Reefs. Being the first part of the geology of the voyage of the Beagle, under the command of Capt. FitzRoy, R.N. during the years 1832 to 1836.* London: Smith, Elder.

1845. *Journal of Researches into the Natural History and Geology of Countries Visited during the Voyage of H.M.S. Beagle round the World, Under the Command of Capt. FitzRoy, R.N.* London: John Murray.

1851–54a. *A Monograph of the Sub-Class Cirripedia.* Vol. I: *The Lepadidae, or Pedunculated Cirripedes*, 1851. Vol. II: *The Balanidae, (or Sessile Cirripedes); the Verrucidae, etc.*, 1854. London: The Ray Society.

1851–54b. *A Monograph of the Fossil Lepadidae, or Pedunculated Cirripedes of Great Britain*, 1851. *A Monograph of the Fossil Balanidae and Verrucidae of Great Britain*, 1854. London: Palaeontographical Society.

1859. *On the Origin of Species by Means of Natural Selection, or the Preservation of Favoured Races in the Struggle for Life.* London: John Murray.

1862. *On the Various Contrivances by which British and Foreign Orchids Are Fertilised by Insects, and on the Good Effects of Intercrossing.* London: John Murray.

1868. *The Variation of Animals and Plants Under Domestication.* London: John Murray.

1871. *The Descent of Man, and Selection in Relation to Sex.* London: John Murray.

1872. *The Expression of the Emotions in Man and Animals.* London: John Murray.

1875. *Insectivorous Plants.* London: John Murray.

1875. *The Movements and Habits of Climbing Plants.* London: John Murray.

1877. *The Different Forms of Flowers on Plants of the Same Species.* London: John Murray.

1881. *The Formation of Vegetable Mould, through the Action of Worms, with Observations on Their Habits.* London: John Murray.

1882. "On the Dispersal of Freshwater Bivalves," *Nature*, vol. 25, April 6, 1882.

Darwin, Charles, and A. R. Wallace. 1858. "On the Tendency of Species to Form Varieties; and on the Perpetuation of Varieties & Species by Means of Natural Selection." Read on July 1, 1858, and first published in the *Journal of the Linnean Society of London*, vol. 3, 1858.

2. Other writings by Charles Darwin, unpublished in his lifetime

Barlow, Nora, ed. 1969. *The Autobiography of Charles Darwin, 1809–1882. With original omissions restored.* New York: W. W. Norton.

Barrett, Paul H., Peter J. Gautrey, Sandra Herbert, David Kohn, and Sydney Smith, eds. 1987. *Charles Darwin's Notebooks, 1836–1844.* Ithaca, NY: Cornell University Press.

Burkhardt, Frederick, ed. 1996. *Charles Darwin's Letters: A Selection.* Cambridge: Cambridge University Press.

———, Sydney Smith, et al., eds. 1985–93. *The Correspondence of Charles*

Darwin, Vols. 1–8, covering 1821–60. Cambridge: Cambridge University Press.

Darwin, Francis, ed. 1887. *The Life and Letters of Charles Darwin, including an autobiographical chapter.* 2 vols. London: John Murray.

———, ed. 1909. *The Foundations of the "Origin of Species": Two Essays Written in 1842 and 1844 by Charles Darwin.* Cambridge: Cambridge University Press (also available on the Web, along with many of Darwin's other writings, edited by John van Wyhe, at http://pages.british library.net/charles.darwin).

———, and A. C. Seward, eds. 1903. *More Letters of Charles Darwin: A Record of his Work in a Series of Hitherto Unpublished Letters.* 2 vols. London: John Murray.

Keynes, Richard Darwin, ed. 1988. *Charles Darwin's Beagle Diary.* Cambridge: Cambridge University Press.

———, ed. *Charles Darwin's Zoology Notes & Specimen Lists from H.M.S. Beagle.* Cambridge: Cambridge University Press.

Stauffer, R. C., ed. 1987. *Charles Darwin's Natural Selection: Being the Second Part of his Big Species Book Written from 1856 to 1858.* Cambridge: Cambridge University Press.

3. Additional sources

Appleman, Philip, ed. 2001. *Darwin: A Norton Critical Edition.* New York: W. W. Norton.

Ayala, Francisco J. 1982. *Population and Evolutionary Genetics: A Primer.* Menlo Park, CA: Benjamin/Cummings.

Baker, Allan J., C. H. Daughterty, Rogan Colbourne, and J. L. McLennan. 1995. "Flightless Brown Kiwis of New Zealand Possess Extremely Subdivided Population Structure and Cryptic Species Like Small Mammals," *Proceedings of the National Academy of Science*, vol. 92.

Barrett, Paul H., Donald J. Weinshank, and Timothy T. Gottleber. 1981. *A Concordance to Darwin's "Origin of Species," First Edition.* Ithaca, NY: Cornell University Press.

Bowlby, John. 1992. *Charles Darwin: A New Life.* New York: W. W. Norton.

Bowler, Peter J. 1989. *Evolution: The History of an Idea*. Berkeley: University of California Press.

———. 1992. *The Eclipse of Darwinism: Anti-Darwinian Evolutionary Theories in the Decades Around 1900*. Baltimore: Johns Hopkins University Press.

———. 1992. *The Non-Darwinian Revolution: Reinterpreting a Historical Myth*. Baltimore: Johns Hopkins University Press.

Brent, Peter. 1983. *Charles Darwin: A Man of Enlarged Curiosity*. New York: W. W. Norton.

Brooks, John Langdon. 1984. *Just Before the Origin: Alfred Russel Wallace's Theory of Evolution*. New York: Columbia University Press.

Browne, Janet. 1983. *The Secular Ark: Studies in the History of Biogeography*. New Haven: Yale University Press.

———. 1996. *Charles Darwin: Voyaging*. Princeton: Princeton University Press.

———. 2002. *Charles Darwin: The Power of Place*. New York: Alfred A. Knopf.

Burkhardt, Richard W., Jr., ed. 1977. *The Spirit of System: Lamarck and Evolutionary Biology*. Cambridge, MA: Harvard University Press.

Calder, William A. III. 1978. "The Kiwi," *Scientific American*, vol. 239.

———. 1979. "The Kiwi and Egg Design: Evolution as a Package Deal," *BioScience*, vol. 29, no. 8.

Camerini, Jane R., ed. 2002. *The Alfred Russel Wallace Reader: A Selection of Writings from the Field*. Baltimore: Johns Hopkins University Press.

Chambers, Robert. 1994 (1844). *Vestiges of the Natural History of Creation, and Other Evolutionary Writings*, ed. James Secord. Chicago: University of Chicago Press.

Cracraft, Joel. 1974. "Phylogeny and Evolution of the Ratite Birds," *Ibis*, vol. 116, no. 4.

Dawkins, Richard. 1991. *The Blind Watchmaker*. London: Penguin Books.

Dennett, Daniel C. 1995. *Darwin's Dangerous Idea: Evolution and the Meanings of Life*. New York: Simon & Schuster.

Depew, David J., and Bruce H. Weber. 1996. *Darwinism Evolving: Systems Dynamics and the Genealogy of Natural Selection*. Cambridge, MA: MIT Press.

————, eds. 1985. *Evolution at a Crossroads: The New Biology and the New Philosophy of Science*. Cambridge, MA: MIT Press.

Desmond, Adrian. 1992. *The Politics of Evolution: Morphology, Medicine, and Reform in Radical London*. Chicago: University of Chicago Press.

————, and James Moore. 1991. *Charles Darwin: The Life of a Tormented Evolutionist*. New York: Warner Books.

Dobzhansky, Theodosius. 1982 (1937). *Genetics and the Origin of Species*. New York: Columbia University Press.

Eldredge, Niles. 1986. *Time Frames: The Rethinking of Darwinian Evolution and the Theory of Punctuated Equilibria*. New York: Touchstone.

————. 1995. *Reinventing Darwin: The Great Debate at the High Table of Evolutionary Theory*. New York: John Wiley & Sons.

————. 2000. *The Triumph of Evolution, and the Failure of Creationism*. New York: W. H. Freeman.

————. 2005. *Darwin: Discovering the Tree of Life*. New York: W. W. Norton.

Ellegård, Alvar. 1990 (1958). *Darwin and the General Reader: The Reception of Darwin's Theory of Evolution in the British Periodical Press, 1859–1872*. Chicago: University of Chicago Press.

Fisher, R. A. 1999 (1930). *The Genetical Theory of Natural Selection: A Complete Variorum Edition*. Oxford: Oxford University Press.

Freeman, R. B. 1977. *The Works of Charles Darwin: An Annotated Bibliographical Handlist*. Hamden, CT: Archon Books.

————. 1978. *Charles Darwin: A Companion*. Hamden, CT: Archon Books.

Fuller, Errol, ed. 1991. *Kiwis: A Monograph of the Family Apterygidae*. Shrewsbury, Shropshire: Swan Hill Press.

Futuyma, Douglas J. 1995. *Science on Trial: The Case for Evolution*. Sunderland, MA: Sinauer Associates.

————. 1998. *Evolutionary Biology*. Sunderland, MA: Sinauer Associates.

Ghiselin, Michael T. 1984. *The Triumph of the Darwinian Method*. Chicago: University of Chicago Press.

Glick, Thomas F., and David Kohn, eds. 1996. *Darwin on Evolution: The Development of the Theory of Natural Selection*. Indianapolis: Hackett.

Godfrey, Laurie Rohde, ed. 1985. *What Darwin Began: Modern Darwinian and Non-Darwinian Perspectives on Evolution*. Boston: Allyn & Bacon.

Goldschmidt, Richard. 1982 (1940). *The Material Basis of Evolution*. New Haven: Yale University Press.

Gosse, Edmund. 1989 (1907). *Father and Son: A Study of Two Temperaments*. London: Penguin Books.

Gould, Stephen Jay. 1977. *Ever Since Darwin: Reflections in Natural History*. New York: W. W. Norton.

———. 2002. *The Structure of Evolutionary Theory*. Cambridge, MA: The Belknap Press of Harvard University Press.

Grant, Peter R. 1986. *Ecology and Evolution of Darwin's Finches*. Princeton: Princeton University Press.

Gray, Asa. 1860. "Darwin on the Origin of Species," *The Atlantic Monthly* (July 1860).

Grzelewski, Derek. 2000. "Night Belongs to the Kiwi," *Smithsonian*, vol. 30, no. 12 (March 2000).

Healey, Edna. 2001. *Emma Darwin: The Inspirational Wife of a Genius*. London: Headline.

Himmelfarb, Gertrude. 1968. *Darwin and the Darwinian Revolution*. New York: W. W. Norton.

Hofstadter, Richard. 1955. *Social Darwinism in American Thought*. Boston: Beacon Press.

Hull, David L. 1983. *Darwin and His Critics: The Reception of Darwin's Theory of Evolution by the Scientific Community*. Chicago: University of Chicago Press.

———. 1988. *Science as a Process: An Evolutionary Account of the Social and Conceptual Development of Science*. Chicago: University of Chicago Press.

———. 1989. *The Metaphysics of Evolution*. Albany: State University of New York Press.

Huxley, Julian. 1943. *Evolution: The Modern Synthesis*. New York: Harper & Bros.

Huxley, Thomas Henry. 1906. *Man's Place in Nature and Other Essays*. London: J. M. Dent & Sons.

Jones, Steve. 2000. *Darwin's Ghost: "The Origin of Species" Updated*. New York: Ballantine Books.

Kaufmann, Stuart. 1995. *At Home in the Universe: The Search for Laws of Self-Organization and Complexity*. New York: Oxford University Press.

Keith, Sir Arthur. 1955. *Darwin Revalued*. London: Watts.

Keynes, Randal. 2002. *Darwin, His Daughter, and Human Evolution*. New York: Riverhead Books (published in the UK as *Annie's Box*).

Kohn, David, ed. 1985. *The Darwinian Heritage*. Princeton: Princeton University Press.

Lack, David. 1968 (1947). *Darwin's Finches: An Essay on the General Biological Theory of Evolution*. Gloucester, MA: Peter Smith.

Larson, Edward J. 1998. *Summer for the Gods: The Scopes Trial and America's Continuing Debate Over Science and Religion*. Cambridge, MA: Harvard University Press.

———. 2001. *Evolution's Workshop: God and Science on the Galápagos Islands*. New York: Basic Books.

Lewontin, Richard. 2000. *It Ain't Necessarily So: The Dream of the Human Genome and Other Illusions*. New York: New York Review of Books.

Litchfield, Henrietta, ed. 1915. *Emma Darwin: A Century of Family Letters, 1792–1896*. 2 vols. New York: D. Appleton.

Lyell, Charles. 1989–90 (1830–33). *Principles of Geology*. Dehra Dun, India: Bishen Singh Mahendra Pal Singh.

Mabey, Richard. 1987. *Gilbert White*. London: Century.

Malthus, T. R. 1992 (1803). *An Essay on the Principle of Population*, ed. Donald Winch. Cambridge: Cambridge University Press.

Marchant, James, ed. 1975 (1916). *Alfred Russel Wallace: Letters and Reminiscences*. New York: Arno Press.

Margulis, Lynn, and Dorion Sagan. 2002. *Acquiring Genomes: A Theory of the Origins of Species*. New York: Basic Books.

Mayr, Ernst. 1964 (1942). *Systematics and the Origin of Species*. New York: Dover.

———. 1976. *Evolution and the Diversity of Life: Selected Essays*. Cambridge, MA: The Belknap Press of Harvard University Press.

———. 1982. *The Growth of Biological Thought: Diversity, Evolution, and Inheritance*. Cambridge, MA: The Belknap Press of Harvard University Press.

———. 1991. *One Long Argument: Charles Darwin and the Genesis of Modern Evolutionary Thought.* Cambridge, MA: Harvard University Press.

———. 2001. *What Evolution Is.* New York: Basic Books.

———, and William B. Provine. *The Evolutionary Synthesis: Perspectives on the Unification of Biology.* 1980. Cambridge, MA: Harvard University Press.

McKinney, H. Lewis. 1972. *Wallace and Natural Selection.* New Haven: Yale University Press.

———, ed. 1971. *Lamarck to Darwin: Contributions to Evolutionary Biology, 1809–1859.* Lawrence, KS: Coronado Press.

McLennan, J. A., M. R. Rudge, and M. A. Potter. 1987. "Range Size and Denning Behaviour of Brown Kiwi, *Apteryx australis mantelli*, in Hawke's Bay, New Zealand," *New Zealand Journal of Ecology*, vol. 10.

Mendel, Gregor. 1965 (1866). *Experiments in Plant Hybridisation,* trans. by the Royal Horticultural Society of London. Cambridge, MA: Harvard University Press.

Moore, James R. 1989. "Of Love and Death: Why Darwin 'Gave up Christianity,'" in James R. Moore, ed., *History, Humanity, and Evolution: Essays for John C. Greene.* Cambridge: Cambridge University Press.

Numbers, Ronald L. 1992. *The Creationists.* Berkeley: University of California Press.

———. 1998. *Darwinism Comes to America.* Cambridge, MA: Harvard University Press.

Ospovat, Dov. 1995. *The Development of Darwin's Theory: Natural History, Natural Theology, and Natural Selection, 1838–1859.* Cambridge: Cambridge University Press.

Otte, Daniel, and John A. Endler, eds. 1989. *Speciation and Its Consequences.* Sunderland, MA: Sinauer Associates.

Owen, Richard. 1838. "On the Anatomy of the Apteryx (*Apteryx australis* Shaw)," *Proceedings of the Zoological Society of London*, vol. 6.

Packard, Alpheus S. 1980 (1901). *Lamarck, the Founder of Evolution: His Life and Work.* New York: Arno Press.

Padian, Kevin. 1999. "Charles Darwin's View of Classification in Theory and Practice," *Systematic Biology*, vol. 48, no. 2.

Pagel, Mark, ed. in chief. 2002. *Encyclopedia of Evolution*. Oxford: Oxford University Press.

Paley, William. 1842 (1802). *Natural Theology: or, Evidences of the Existence and Attributes of the Deity, Collected from the Appearances of Nature*. Boston: Gould, Kendall & Lincoln.

Palumbi, Stephen R. 2001. *The Evolution Explosion: How Humans Cause Rapid Evolutionary Change*. New York: W. W. Norton.

Peckham, Morse, ed. 1959. *The Origin of Species by Charles Darwin, A Variorum Text*. Philadelphia: University of Pennsylvania Press.

Provine, William B. 1986. *Sewall Wright and Evolutionary Biology*. Chicago: University of Chicago Press.

Raby, Peter. 2001. *Alfred Russel Wallace: A Life*. Princeton: Princeton University Press.

Raverat, Gwen. 1953. *Period Piece*. New York: W. W. Norton.

Reid, Brian, and G. R. Williams. 1975. "The Kiwi," in G. Kuschel, ed., *Biogeography and Ecology in New Zealand*. The Hague: Dr. W. Junk.

Ridley, Mark. 1993. *Evolution*. Cambridge, MA: Blackwell Scientific Publications.

Ridley, Matt. 2000. *Genome: The Autobiography of a Species in 23 Chapters*. New York: HarperCollins.

Ross, Herbert H. 1974. *Biological Systematics*. Reading, MA.: Addison-Wesley.

Secord, James A. 1981. "Nature's Fancy: Charles Darwin and the Breeding of Pigeons," *Isis*, vol. 72: 262.

———. 2000. *Victorian Sensation: The Extraordinary Publication, Reception, and Secret Authorship of "Vestiges of the Natural History of Creation."* Chicago: University of Chicago Press.

Short, Keir. 1837. "Remarks Upon the Apteryx," *Proceedings of the Zoological Society of London*, vol. 5.

Smith, Charles H., ed. 1991. *Alfred Russel Wallace: An Anthology of His Shorter Writings*. Oxford: Oxford University Press.

Smith, John Maynard. 1993 (1958). *The Theory of Evolution*. Cambridge: Cambridge University Press.

Stott, Rebecca. 2003. *Darwin and the Barnacle*. New York: W. W. Norton.

Sulloway, Frank J. 1982. "Darwin and His Finches: The Evolution of a Legend," *Journal of the History of Biology*, vol. 15, no. 1.

Turrill, W. B. 1963. *Joseph Dalton Hooker: Botanist, Explorer, and Administrator*. London: Scientific Book Club.

Wallace, Alfred Russel. 1852. "On the Monkeys of the Amazon," *Proceedings of the Zoological Society of London*, vol. 20, reprinted in Camerini, ed. (2002).

———. 1855. "On the Law which has Regulated the Introduction of New Species," *Annals and Magazine of Natural History*, vol. 16, reprinted in Wallace (1969).

———. 1858. "On the Tendency of Varieties to Depart Indefinitely from the Original Type," read aloud to the Linnean Society on July 1, 1858; first published as part of Darwin and Wallace (1858); reprinted in Weigel, ed., "On the Tendency of Species . . ." (1975), and in Wallace (1969).

———. 1889. *Travels on the Amazon and Rio Negro*. New York: Ward, Lock.

———. 1905. *My Life: A Record of Events and Opinions*. 2 vols. London: Chapman & Hall.

———. 1962 (1869). *The Malay Archipelago*. New York: Dover.

———. 1969 (1870, 1878). *Natural Selection and Tropical Nature: Essays on Descriptive and Theoretical Biology*. Westmead, Surrey: Gregg International.

Wedgwood, Barbara, and Hensleigh Wedgwood. 1980. *The Wedgwood Circle, 1730–1897: Four Generations of a Family and Their Friends*. Westfield, NJ: Eastview Editions.

Weigel, Robert D., ed. 1975. "On the Tendency of Species to Form Varieties; and On the Perpetuation of Varieties & Species by Natural Means of Selection," by Charles Darwin and A. R. Wallace. Bloomington: Scarlet Ibis Press.

Weiner, Jonathan. 1994. *The Beak of the Finch: A Story of Evolution in Our Time*. New York: Alfred A. Knopf.

Wesson, Robert. 1993. *Beyond Natural Selection*. Cambridge, MA: MIT Press.

White, Michael, and John Gribbin. 1997. *Darwin: A Life in Science*. New York: Plume/Penguin.

Williams, George C. 1974 (1966). *Adaptation and Natural Selection: A Critique of Some Current Evolutionary Thought.* Princeton: Princeton University Press.

Wilson, Leonard G., ed. 1970. *Sir Charles Lyell's Scientific Journals on the Species Question.* New Haven: Yale University Press.

Worster, Donald. 1985. *Nature's Economy: A History of Ecological Ideas.* Cambridge: Cambridge University Press.

Wrigley, E. A., and David Souden, eds. 1986 (1826, 1803). *The Works of Thomas Robert Malthus.* Vol. 2, *An Essay on the Principle of Population, The sixth edition (1826) with variant readings from the second edition (1803).* London: William Pickering.

Acknowledgments

This book had its origin, seven years ago, in an invitation from James Atlas: Would I like to do Charles Darwin for his forthcoming series of compact biographies, to be known as the Penguin Lives? When I answered, by way of demurral, that Darwin had been well served in long, authoritative, excellent biographies within the past decade, I was thinking of Janet Browne's work, and of Desmond and Moore's. Never mind, said Jim, explaining what sort of book he wanted: radically concise, essayistic, and writerly more than scholarly. Those big biographies aren't your competition, he said, they're your resources. Having agreed, I was delayed by other projects until, by the time my desk was cleared for Darwin, Jim had parted with Penguin and founded a new series, Great Discoveries, at W. W. Norton. So I moved the book there, partly because Jim was its godfather and partly because Norton was already my primary publishing home.

I thank Jim—for his large vision of small biographies, and for the trust behind his invitation. Thanks also to Jesse Cohen, of Atlas Books, who has done much for this book in particular and for the Great Discoveries series within which it appears. At Norton, my longtime editor Maria Guarnaschelli has been a vital partner again, with her keen editorial insights and impassioned support. Erik Johnson and Robin Muller, as Maria's assistants, have helped me with many details. I'm grateful to everyone at Norton for their various contributions. Carolyn Carlson was a welcoming contact at Penguin while the book was lodged there. Renée Wayne Golden, my agent, played her usual crucial role, reconciling the possible with the necessary.

Michael E. Gilpin has been, for almost twenty years, my friendly consulting expert on the biological sciences. There's no better person with whom to spend time on mountain bikes or skis while discussing the fine points of theoretical population biology. Mike read this book in draft, as he has my last several, and offered valuable feedback. I'm also grateful to three other readers whose expertise and detailed comments helped me avoid many (although probably not all) mistakes and distortions: Kevin Padian, Michael Reidy, and Stan Rachootin. Obviously, they're not answerable for the flaws of the final product. Michelle Harris was my vigilant professional fact-checker. Ann Adelman did the rigorous but restrained copyedit. I'm grateful to many other people, not all listed here, for various acts of support and incitement, but I'll mention some specifics. Prosser Gifford invited me to deliver a Bradley Lecture on *The Origin of Species* at the Library of Congress, which served as my point of entry into the book project. Bill Allen and Oliver Payne, of *National Geographic* magazine, provided indirect support by assigning me to write a feature story on the evidence for evolution, published (November 2004) as "Was Darwin Wrong?", and many other people at *National Geographic*, notably Bernard Ohanian and Mary McPeak, helped bring the story to print. Douglas Futuyma, Philip Gingerich, Niles Eldredge, Ian Tattersall, and Eugenie Scott gave generously of their time and ideas during my research for that article. Joan and Arnold Travis arranged for me to be hosted on a return trip, after seventeen years, to the Galápagos. Dennis Hutchinson gave me a very crucial little book containing the 1858 papers. David Singel advised me on the chemistry of prussic acid. Matt Ridley alerted me to the connection between W. D. Crick and his grandson Francis. Mary and Will Quammen were helpful, as ever, in immeasurable ways, and Betsy Gaines Quammen lit the home in which this book was written. I also owe a glad debt of thanks to the late R. W. B. Lewis and his wife Nancy Lewis, for literary and personal generosities going back forty years.

Index

6/08 - ⑦ √7/08